A LITTLE BOOK
ON
WATER SUPPLY

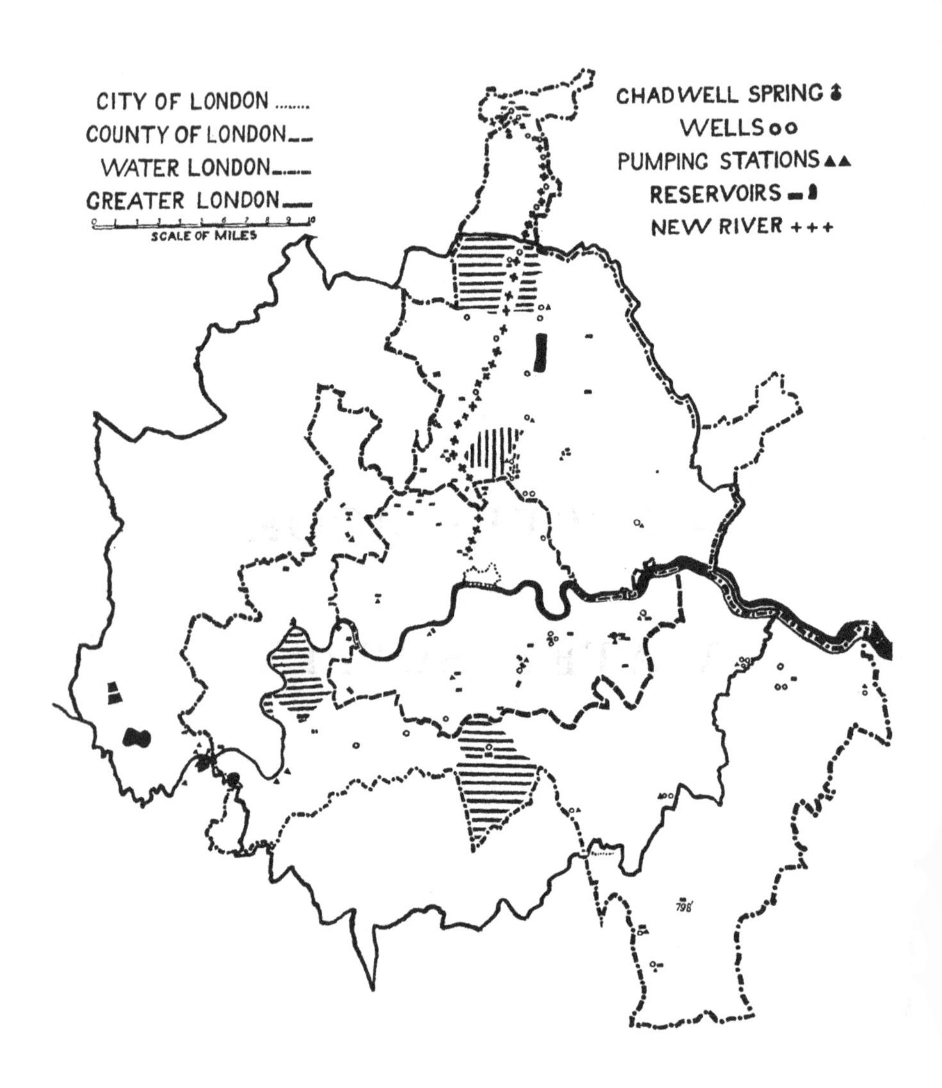

CITY OF LONDON
COUNTY OF LONDON
WATER LONDON
GREATER LONDON
SCALE OF MILES
CHADWELL SPRING
WELLS
PUMPING STATIONS
RESERVOIRS
NEW RIVER

A LITTLE BOOK
ON
WATER SUPPLY

BY

WILLIAM GARNETT, M.A., D.C.L.
Formerly Fellow of St John's College, Cambridge

CAMBRIDGE
AT THE UNIVERSITY PRESS
1922

CAMBRIDGE
UNIVERSITY PRESS

University Printing House, Cambridge CB2 8BS, United Kingdom

Published in the United States of America by Cambridge University Press, New York

Cambridge University Press is part of the University of Cambridge.

It furthers the University's mission by disseminating knowledge in the pursuit of education, learning and research at the highest international levels of excellence.

www.cambridge.org
Information on this title: www.cambridge.org/9781107650480

First published 1922
First paperback edition 2014

A catalogue record for this publication is available from the British Library

ISBN 978-1-107-65048-0 Paperback

TO

MICHAEL

PREFACE

THIS little book when in manuscript was condemned by a very high authority on educational publications because it did not enable the reader "to prepare for any specific examination." Students whose sole object is to pass an examination are warned against devoting time to a book which is intended to give them only some general knowledge respecting the sources and mode of supply of one of the chief necessaries of life, that, in fact, which for all time has determined the localities which men have selected for residence, and at the same time to arouse an interest in a number of sciences which have an important bearing on everyday life.

Its origin was a quantity of information very carefully prepared by Mr G. P. Warner Terry, Barrister-at-Law, F.A.A., the Statistical Officer of the Metropolitan Water Board and Secretary of the Municipal Waterworks Association. This matter was sent by the Water Board in 1914 to the Education Committee of the London County Council together with a list of illustrative lantern slides belonging to the Board and available for lecture purposes. These slides were offered on loan to the Council for the use of any schools within its area. A commencement was made at the time by the present writer in the preparation of descriptive matter to be sent to the schools with the slides. The outbreak of war prevented any immediate action being taken on this proposal. The fear of injury to the water supply was illustrated by the action of the scout who called the attention of the special constabulary to a fisherman in the neighbourhood of a storage reservoir with a creel marked "POISSON." But while it was deemed undesirable during the war to publish information respecting London waterworks, the descriptive pamphlet gradually assumed the dimensions of a school reading-book; and it was thought that some account of public water supply would not only be serviceable to teachers and students in creating an interest in the proposed lectures, but would also be a useful addition to general education, especially in the new Part-Time Continuation Schools. Moreover there is a special need for a book on this subject in Training Colleges, and particularly in those Colleges which are attended by students of Domestic Economy and Hygiene. From the point of view of general education, water supply has a bearing on a great many school subjects and the teacher can use this little book as dealing with a central interest and affording a series of texts for lessons on the greater portion of the "Circle of the Sciences," including mensuration, mechanics, hydrostatics and hydraulics, surveying, geography, geology, heat and heat engines, chemistry, bacteriology, pathology and even electric distribution of power from the "white coal" of the mountains. Short accounts have been introduced of the water supply of Manchester, Glasgow, Liverpool, Birmingham, New York and the ancient city of Jerusalem; but London affords examples of nearly all the ordinary sources of supply except the upland sources, which have not yet been drawn upon for "Water London." For details of collection and distribution no other English city can afford such variety as the Metropolis, and the experience of the year 1921 suggests that the time is not far distant when an upland supply will be required by the Water Authority of London.

When the manuscript was nearly complete it was offered to the London County Council if the Education Committee desired to issue it as a Council publication; but the Council's representatives wisely distinguished between public water supply and a school-book descriptive thereof, holding that while the one should be in the hands of the Local Authority the other might safely be left to commercial enterprise. This decision accounts for the publication of the book in its present form.

The author makes no claim to originality. He has not discovered new sources of water supply or invented new methods of purification or distribution. He has simply attempted to present, in a form convenient for teachers and pupils, facts which are well known but not conveniently accessible, and he is unaware of any other school-book in which information of the same type is collected.

For most of the particulars respecting the recent Water Companies of London and the present Water Board, and its work, and for much of the early history of London's water supply the author is indebted to the Statistical Officer and other officers of the Metropolitan Water Board, especially the District Engineer and Mr H. W. Carmen of the Statistical Officer's Department. *The Springs, Streams and Spas of London* by Alfred Stanley Ford has been freely consulted and some material has been taken from *Studies in Water Supply* by Sir Alex. Houston, the Water Examiner to the Board. The particulars of the artesian wells of London and the water gradients in the chalk were taken from the *Records of London Wells* by G. Barrow and L. J. Wills, published by the Geological Survey. The account of the making of the New River is partly taken from Smiles' *Lives of the Engineers*, while other matter has been derived from the article on "Water Supply" in the *Encyclopaedia Britannica* by the Engineer of the Liverpool Corporation and from other articles in the same encyclopaedia, and to the publishers of the encyclopaedia the author is indebted for some illustrations. The account of the ancient water supply of Jerusalem is taken from the work of Principal Sir George Adam Smith. The author's thanks are also due to the officers of the Manchester City Council for figures respecting the supply from Thirlmere and to Professor Hewlett, of King's College, London, for the illustrations of bacteria; to the Director of the Geological Survey of England and Wales, the Director of the Meteorological Office and the Controller of His Majesty's Stationery Office for permission to make extracts from State publications; and to Messrs Isler and Company for the use of blocks relating to deep well pumping.

The author also wishes to acknowledge his great indebtedness to the staff of the Cambridge University Press for their advice and suggestions and for the efficient manner in which they have dealt with his MS. and sketches.

The ornament on the title page is the old seal of the New River Company and indicates the ultimate source of all water supply. The legend is taken from the Vulgate of Amos iv. 7—ET PLUI SUPER UNAM CIVITATEM.

W. G.

December 29th, 1921

BIBLIOGRAPHY

The following are a few books among many which, in addition to those mentioned in the preface and text of this book, may well be consulted by teachers and others who desire further and more first-hand information on Water Supply.

The Several Memoirs of the Geological Survey on the Water Supply of the English Counties. (H. M. S. O.)

The Annual Reports of the Metropolitan Water Board. (P. S. King and Son.)

The Geology of Water Supply. H. B. WOODWARD. (Edward Arnold.)

The London Water Supply. R. SISLEY. (The Scientific Press.)

The Water Supply of Towns. W. K. BURTON. (Crosby, Lockwood and Son.)

Rainfall Reservoirs and Water Supply. Sir ALEX. BINNIE. (Constable.)

Conveyance and Distribution of Water for Water Supply. E. WEGMANN. (Van Nostrand, New York, and Constable.)

Water Supply. W. P. MASON. (Chapman and Hall.)

The Examination of Water and Water Supplies. J. C. THRESH. (J. and A. Churchill.)

Well Boring. C. ISLER. (Spon.)

History and Description of the Thirlmere Water Scheme. Sir J. J. HARWOOD. (Blacklock and Co., Manchester.)

The Thirtieth Annual Report of the Local Government Board—Supplement on Lead Poisoning and Water Supplies. (H. M. S. O.)

On the Structure of Moving Cyclones. J. BJERKNES. (Geofysiske Publikationer, Vol. I, No. 2. Grondahl and Sons, Kristiania.)

Meteorological Conditions for the Formation of Rain. J. BJERKNES and H. SOLBERG. (Geofysiske Publikationer, Vol. II, No. 3. Grondahl and Sons, Kristiania.)

CONTENTS

A LITTLE BOOK ON WATER SUPPLY

Nature's Water Supply

Everybody knows that a good supply of sufficiently pure water is essential to life and that in towns water is required for a great many purposes besides drinking and washing. Apart from the water necessary for special manufacturing purposes something like 40 gallons a day have to be supplied for each person in large towns for domestic, trade and public purposes. This means that the amount of water which has to be supplied every year to the 117 square miles of the Administrative County of London is nearly seventy thousand million gallons. If a wall were built all round the square mile of the City of London to the height of St Paul's the tank so formed would not contain a year's supply of water for the Administrative County; and the Greater London which is supplied with water by the Metropolitan Water Board is very much larger than the Administrative County and covers an area of 559 square miles with a population of 6,930,000.

The great storehouse of water is the ocean, but sea water contains far too much common salt and other solids in solution to be fit for drinking, washing or use in boilers, and for the majority of people, including Londoners, the sea is a long way off. If you take a little sea water in a glass and leave it for a few days, the water will dry up and leave some white salt. The salt will taste like ordinary table salt but slightly bitter, because it contains some Epsom salts as well as common salt; but what has become of the water? It has passed into the air, leaving the salt behind. It has evaporated, or become vapour, but it is only pure water that has gone into the air. If we could get it back again it would be quite free from salt. The whole surface of the ocean is giving water to the air continually, just as the water escaped from the glass, but the amount of water which can remain as vapour in a cubic foot of air (or empty space) depends entirely on the temperature. The warmer the air the greater the amount of vapour, and if water is kept at the boiling-point it will blow away the air altogether and fill the whole space with steam. If air which contains as much vapour as it can is cooled the vapour condenses into tiny drops forming a cloud. If a jug containing very cold water is brought into a warm room full of people drops of water will presently be seen on the outside of the jug. This water has come out of the air. The vapour has been condensed on the cold jug. So when warm air from the surface of the sea is cooled by ascending to a great height, or by mixing with cold air, or by blowing against a cold mountain peak, the water vapour is condensed and a cloud or mist is formed. On a clear, warm day when a strong wind is blowing

Fig. 1. Cervin (the Matterhorn) smoking his pipe.

from Italy against the very cold peak of the Matterhorn some of the moisture, carried with the wind is condensed, forming a beautiful white cloud in the sunshine. As the air passes away from the mountain and mingles with other air which has not been cooled against the snowy peak, while the sun shines on the cloud, it gets warm again and presently, perhaps half a mile away from the mountain, the cloud evaporates and the air becomes clear and the sky blue again. The result is that a cloud remains apparently stationary against the top of the mountain and extending for perhaps half a mile on the leeward side, never being blown away although the wind may be blowing very hard. Of course the cloud is really being carried with the wind, but fresh cloud is always being formed against the mountain and it evaporates as soon as it gets far enough away. The cloud is therefore always changing though it appears to be quite still. Fig. 1 shows the Matterhorn "smoking his pipe." The cloud is a good illustration of condensation and evaporation going on simultaneously in regions a little distance apart. The figure shows that it is formed on the Italian face of the mountain and blown over the rounded peak.

For those who have not studied evaporation and condensation in connection with lessons on heat the following are a few points worth noticing. For example, water will continue to evaporate until the space above it contains a certain amount of vapour which increases rapidly with the temperature and does not depend on the presence or absence of air in the space above the water, though the presence of air renders the evaporation much slower than it would be without the air. When the space contains all the vapour it can at the existing temperature it is said to be saturated. If the vapour in a saturated space is cooled it will condense into drops of water, forming cloud. If you wear spectacles and come into a warm room on a cold day from the open air your spectacles will become dim with dew. If two lots of air, one warmer than the other but both full of vapour, are mixed, the warm air is cooled and the cool air is warmed, but the whole space at the middle temperature cannot hold as much vapour as the two spaces, one hot and the other cold, because as the temperature increases the amount of vapour which any space can hold increases faster and faster. Consequently, if a cold wind and a warm wind, both full of vapour, mix, some of the vapour will be condensed and a cloud will be formed. As we ascend in the atmosphere in a balloon or by going up a hill or mountain the pressure gets less and less. If we carry with us a large bottle of air with a tube passing through the cork closed by a stop-tap and we open the tap when we have gone up a thousand feet or so some of the air will escape, because under the smaller pressure the air expands. As it expands it will get cooler because it does work in lifting the air outside the bottle. If the air is quite full of vapour the cooling will make some of the vapour condense and form a cloud in the bottle. The cooling of warm air as it ascends and expands in the atmospheric circulation is the principal cause of the formation of cloud.

When the little drops of water forming a cloud come together to make larger drops, rain is produced. As a raindrop falls through the cloud it keeps on picking up more and more little drops so that it becomes bigger and bigger as it falls. If the cloud is so cold that the water it contains is frozen snow is formed instead of rain. When the rain falls on the ground some of it generally finds its way into streams and rivers and in course of time again reaches the sea carrying with it some parts of the soil which it has dissolved. This completes the circuit. The sea water evaporates into the air, the air carries it away with the winds perhaps into higher latitudes where the climate is much colder or perhaps on to a cold mountain-side, or perhaps it is carried high up into the atmosphere with a rising current, or possibly it meets with a wind from a cold region; in all these cases some of the vapour is condensed into cloud and may fall as rain or snow and then it may return to the sea by streams and rivers. If the cloud falls as snow on high mountains it may help to make a glacier, or river of ice, and then it may be hundreds of years before it reaches the sea, with the very slow descent of the glacier into regions where the ice will melt; or it may fall on some ice-covered country like Greenland and slide slowly with a sheet of ice down to the sea and then break off into icebergs.

But not all the rain falling on the land makes its way directly to the sea. Very much depends on the kind of soil on which it falls and on the vegetation. Some of the rain will soak a little way into the ground or lie on the surface and evaporate again in course of time, just as it would evaporate from the sea, and so form cloud and rain again. Some will be sucked up by plants and trees and evaporate into the air from their leaves. Some will soak through the porous soil until it reaches clay or hard rock which is waterproof and it will then flow over the surface of the waterproof rock until it reaches some valley where it appears as a spring if it has originally fallen on high land, and then finds its way into a stream or river and so to the sea. Sometimes a porous rock like the Upper and Middle Chalk, the Greensand or the New Red Sandstone, will extend to a thickness of some hundreds of feet under the surface. The rain water will soak into this porous rock and flow out into some distant valley where the lower part is exposed, but if the porous soil extends below sea-level the water will remain in the soil until it is filled up to sea-level, and this water cannot escape unless the water in the rock is raised from wells by pumps or otherwise. We shall have a good deal to say presently about the great reservoir of water in the chalk extending to a depth of 500 or 600 feet or more under London.

Where, as in most of the Welsh mountains, there is only a thin layer of peaty soil lying on waterproof rock, like the slate of the Silurian formation, the rain which does not evaporate and is not absorbed by vegetation soon finds its way down the hillside close to the surface and flows into the rivers and lakes, perhaps a little stained with peat or containing a little acid from the fermentation of the peat, but otherwise almost pure, with hardly any lime or other mineral in solution. In the Thames valley the soil is, as a rule, much more porous than on the Welsh hills. This is true of the oolitic rocks in the upper valley of the Thames as well as of the chalk of the Chilterns and lower Thames valley. While some of the rain, especially after a heavy fall, finds its way to the river in streams along the surface, a good deal of the rain soaks into the soil and reaches the river, after flowing for a long way underground, rising in springs in the river-bed. This water will dissolve some portion of the soil through which it flows and reach the river with common salt and carbonate of lime (chalk) and other mineral substances in solution.

Sometimes a few feet of sandy clay, or brick earth, or clay interspersed with thin seams of sand lie just below the surface but rest on a hard waterproof clay like the London Clay or the boulder clay of the north of England. Rain water can soak into the sandy clay or seams of sand and flow slowly through them but it cannot penetrate the clay below. If a well is dug through the porous soil and a little way into the hard clay water will slowly flow into the well and fill up the watertight hole in the stiff clay. This is an example of a shallow well. Some of the rain falling on the Bagshot sands on Hampstead Heath used to find its way through seams of sand above the London Clay to a well in Well Walk. This was the origin of the Hampstead Pump Room. The water dissolved some iron from the sands in its course. Here and there on the lower parts of the Heath water may still be seen flowing from the sand seams in the drift clay having found its way through them from the Bagshot sands at the top. Water from shallow wells in a thickly populated neighbourhood is not safe for drinking, and most of the water which used to supply the Hampstead well now flows into the sewers through pipes constructed to drain the foundations of the houses.

When the whole soil of a district is porous for a considerable depth, say a few hundred feet, the rain water soaks in and saturates the porous soil up to a certain depth depending on the levels at which it can find its escape into some neighbouring valley, on the distance of the valley and the resistance offered by the rock to the flow of water. It is only with great difficulty that water can flow through sand or chalk, and consequently a mile or two away from the valley the water will stand at a level much above the outflow. If a well is sunk into this porous soil below the level at which the water stands water will collect in the well up to that level and can be drawn out by a bucket or pump. In this case the water is held in the well not because it is sunk into a waterproof stratum but because it is sunk below the level at which the water stands in the porous soil, through which it is constantly flowing. Examples of water levels (contours) in the chalk in the London area are given in the map, p. 64, and described on pp. 64 and 65.

Rain, Well, and River Waters

If rain water be collected in a watch-glass in the country and allowed to evaporate it will dry up leaving scarcely any stain behind it. Pure rain water contains no solids in solution. Water flowing off the surface of slate hills will perhaps leave a little brown residue of peaty matter which will be nearly all destroyed if heated in air to redness. Water taken from a shallow well made in gravel or sandy clay will probably leave a little oxide of iron with some vegetable matter. If the water comes from a shallow well in chalk soil it will leave behind it a quantity of chalk but not nearly so much as the common salt left by the evaporation of the same quantity of sea water. If water from a chalk well is boiled it will become slightly milky and deposit a white sediment of chalk. This forms "fur" in kettles and incrusts steam boilers so that it sometimes has to be removed with a hammer and chisel. The reason why the chalk is thrown out of solution by boiling is that pure water can dissolve very little chalk. It is only when there is free carbonic acid in the water that the chalk is soluble. Soda water would dissolve a good deal of chalk, because it is highly charged with carbonic acid. When the water is boiled the free carbonic acid is driven away and the dissolved chalk is thrown out of solution. If quicklime is added to water containing dissolved chalk the lime seizes on the free carbonic acid and fresh chalk is formed, while the loss of the carbonic acid throws the dissolved chalk out of solution and both lots of chalk fall to the bottom. This system of removing dissolved chalk from water required for steam boilers is often used. It is also sometimes used for purifying drinking water (see p. 82).

It is a curious fact that water from the deep wells in the chalk under London, which will be described later on, contains much less chalk than water from shallow chalk wells. Perhaps this is due to the water deep down in the soil being unable to get the necessary carbonic acid from the air or the soil to enable it to dissolve the chalk and to its having lost its carbonic acid to the iron in passing through the Thanet sands.

When river water is evaporated the residue will depend on the sources from which the river derives its water. If the water has flowed off the surface of hard slaty rocks it will leave very little residue; if it has flowed over farm lands it will bring with it a lot of animal and vegetable matter, some salt and probably nitrates. If it has percolated through sandy soil it will contain a little silica (the basis of sand) and probably some iron. If it has soaked through chalky soil, or even flowed over the surface of chalk, it will dissolve some of the chalk and deposit carbonate of lime when it is evaporated. As most minerals are, to some extent, soluble in water there are very few substances in the earth's crust which may not be found dissolved in the water of one river or another. When the river obtains part of its water from springs, the water of which has travelled for a long way through the ground, it will usually have a lot of mineral matter in solution and resemble spring water. This is the case with the Thames water in a dry summer, for the main supply of the river then comes from springs of water which has flowed through chalk or oolite. Medicinal springs generally contain a considerable amount of mineral salts in solution which the water has dissolved in its journey through the earth.

Hard and Soft Water

If we try to dissolve soap in water which contains chalk in solution the fatty acids in the soap combine with the lime in the chalk forming a white curd and the soap is destroyed so that it will not make a lather or help to remove grease. It is only after all the chalk has combined with the acids that the soap begins to be of any service. A water which destroys soap in this way is called "hard." Rain water and distilled water are soft. Soap lathers in them at once. You have only to wash your hands with soap in chalk water and rain water to feel the reason for calling these waters hard and soft. Water from the Welsh hills and the Cumberland lakes is soft. Water from the chalk or from limestone is hard. Sea water is very hard. Try to wash your hands with soap in a basin of sea water. Hard water, like a smoky atmosphere, is very costly to a town. It necessitates the use of so much extra soap and it furs up kettles and boilers and so wastes coal. On the other hand, water which contains no lime (or chalk) is not so good for drinking purposes, especially for children who require lime for the growth of bone and teeth. Upland water containing acid from fermenting peats, as it flows from some high moorlands, also has the power of dissolving a small amount of lead oxide if it passes through a great length of lead pipe or is stored in lead-lined cisterns. Hence in some towns which are provided with a supply of very soft water from moorland sources iron service pipes are used instead of lead to avoid lead poisoning. Water containing a little lime will neutralise any trace of vegetable acid and the neutralised water will not attack lead. Lead pipes may safely be used for any water supplied in London.

To make distilled water, ordinary water is boiled and the steam condensed. The process is exactly like the formation of rain from the sea, but is carried on at a higher temperature to save time.

Gases and Organic Impurities in Natural Waters

Rain water absorbs from the air a little oxygen, nitrogen and carbonic acid. In percolating through soil which contains decomposing animal or vegetable matter it will take up more carbonic acid, for water at ordinary temperature will dissolve about its own volume of carbonic acid at any pressure to which it may be exposed. If the water is saturated with the gas and the pressure is relieved the gas expands, and the water sparkles or effervesces. This is seen in soda water, lemonade, gingerbeer, etc., which have been charged with carbonic acid at a high pressure produced by pumps. It is also seen in champagne, "stone gingerbeer" and other liquids which have been allowed to ferment and produce carbonic acid from sugar when corked up in bottles. Some natural waters get charged with carbonic acid underground at a high pressure and effervesce when discharged from the spring. These form sparkling natural mineral waters. When water from a shallow well sparkles it is generally an indication that it has been contaminated by decomposing animal or vegetable matter in the soil and is dangerous to drink. If water from the tap is heated in a glass beaker or flask it will be generally seen that bubbles of gas are formed, and rise to the surface long before the water begins to boil. These bubbles are the oxygen and nitrogen (air) and carbonic acid or other gases in solution, and they carry with them more and more water vapour as the temperature rises, until, on reaching the boiling-point, they consist of almost pure steam.

In addition to the gases and mineral solids in solution, such as salts of iron, lime and magnesium and common salt, water which has percolated through the top soil only and then found its way into shallow wells or rivers will contain animal and vegetable matter in solution or suspension. In course of time this organic matter in a running stream becomes oxidised, or burnt up, by means of oxygen absorbed from the air, and this process is helped very much if in the course of the stream there are waterfalls or other obstructions which break up the water and enable it to entangle and dissolve the oxygen from the air. When the organic matter has been destroyed in this way it will generally, especially if it is animal matter, leave its traces behind in the water in the form of combined nitrogen, either as ammonia or nitric acid, and the quantity of combined nitrogen found in water used commonly to be taken as an indication of the extent to which it had been contaminated by organic matter since it fell as rain (previous sewage contamination).

The organic matter itself would be comparatively harmless, but it is generally accompanied by living organisms, commonly known as microbes, which derive their food from the animal or vegetable matter in the water and which would not live long or multiply if the water contained no food. It is largely due to these microbes that the organic matter is ultimately fermented and destroyed, so that most of them may be regarded as friendly and quite harmless, but on the other hand some microbes are the cause of fevers and other diseases. The disease germ most likely to be found in rivers or shallow wells is that which causes enteric, or typhoid, fever, but cholera germs may also be distributed with the water supply. The typhoid bacilli or cholera vibrios, as the microbes which cause these diseases are called, cannot be seen by the naked eye but are visible with the help of a very high-power microscope. It is the danger of

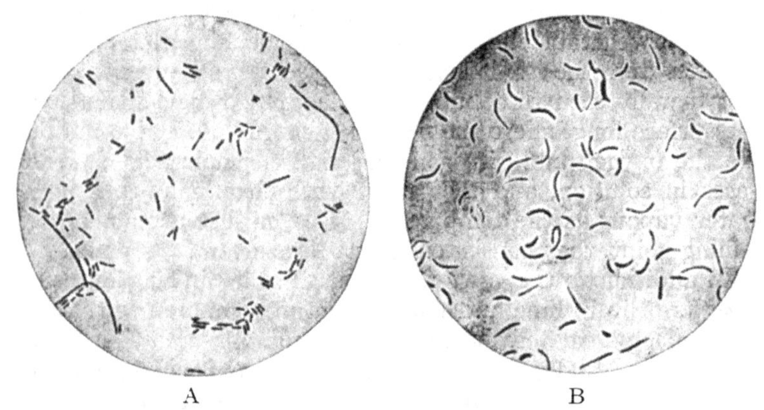

Fig. 2. A. Typhoid bacilli × 1500. B. Cholera vibrios (comma bacillus) × 1500.

their presence which makes it so important for the public authorities to keep a very careful watch over the water which is used for drinking purposes. Pathogenic microbes, or disease generating microbes, are usually killed by boiling the water for a few minutes, so that if there is any suspicion of the purity of drinking water the best course is to boil it, although the loss of the dissolved air and carbonic acid gas makes boiled water taste very "flat." Some aerated table waters are made from distilled water. As all this water has been converted into steam and then condensed, it necessarily leaves behind in the boiler the mineral and non-volatile organic matter, including the germs, with which it may have been contaminated. It is afterwards aerated artificially.

The microbes common in river water have a habit of destroying pathogenic microbes. The vibrio of Asiatic cholera is a very delicate organism quite unable to survive for long the rough-and-tumble of life in an English river.

When water, after passing through the surface soil, penetrates to a very great depth through chalk or sandstone or other porous rock and is then pumped from deep wells it is free from microbes and from organic matter, mainly on account of the filtering action of the rock but partly, no doubt, because of the long time which must elapse before the water reaches the well after it has passed through the surface soil. If impure water is stored for only a month in an open reservoir and not subjected to any filtering action it becomes much less impure, for not only does the heavy solid matter in suspension sink to the bottom but the organic matter is gradually used up by fermentation or oxidation and the microbes, finding their food supply fail, die, so that the number of microbes in a gallon of river water which has stood for a month in a storage reservoir is very small compared with the number in the same quantity of water taken directly from the river. The extent of the reduction depends on the temperature and on the amount of germ food in the water, but the microbes of typhoid and cholera are all killed with less than a month's storage (see p. 82).

Storm Waters

Although the water from springs and deep wells may be sufficiently pure for drinking without any further process of purification, water taken from rivers or shallow wells, which receive the rain when it has flowed only over the surface of the land or percolated through the top soil, must be purified before it can be supplied for drinking purposes. A river like the Thames after a very heavy rainfall is filled with water which has poured off the land and carried with it the washings from the roads and some of the manure from the fields. When the river is thus in flood the water is dark and turbid and quite unfit for domestic use. Two or three days after the heavy rainfall has ceased, the river will be flowing quite strongly but the water will be clear and may be collected, to be used for public supply after purification. The quantity of water flowing down the Thames above the tidal limits varies enormously with the time of year and the recent condition of the weather. On November 18th, 1894, the flow at Teddington Weir, below the Water Companies' intakes, amounted to 20,136,000,000 gallons in one day, but on September 11th, 1896, it was only 23,000,000 gallons, so that on the former date the flow was nearly a thousand times as great as on the latter. In fact, the smaller flow was very much less than the average amount which is drawn from the river in one day for the supply of part of London. We shall see presently how the difficulty caused by this great variation in the supply is overcome by means of storage reservoirs.

If the reader has spent a summer holiday at the seaside where the drains from the village or town run out into the sea, and in a part of the country where the roads are made up with red sandy gravel, he will probably have noticed that after a heavy thunderstorm the sea for some distance in front of the end of a drain is no longer blue or green but of a dull yellow colour. This is due to the sand and other soil which the heavy rain has washed from the roads and carried into the drain. The same kind of action takes place when the surface drainage runs into a river, and it is obvious that the river is then in no condition for supplying water for domestic use.

Aqueducts

Throughout the world's history as large centres of population have developed it has usually been necessary to seek a water supply at a distance, as before the days of steam pumps it was necessary that the source of supply should be at such an elevation that the water would flow naturally from the source to the city if a suitable conduit were provided. A source might be plentiful and near but if it were at a lower level than the city the water could not be utilised except by the expenditure of animal (including human) power, and the water carrier became a necessary institution. The artificial channel along which the water is conveyed is an aqueduct, and both Greeks and Romans constructed these artificial channels for great distances in order to provide a copious supply of pure water. In the earliest days the aqueduct was an open channel like a river. The writer remembers when the New River was an open channel with green banks through Hornsey, Highbury, Islington and into Clerkenwell. In order that the water might flow steadily in such a channel it was necessary that there should be a tolerably

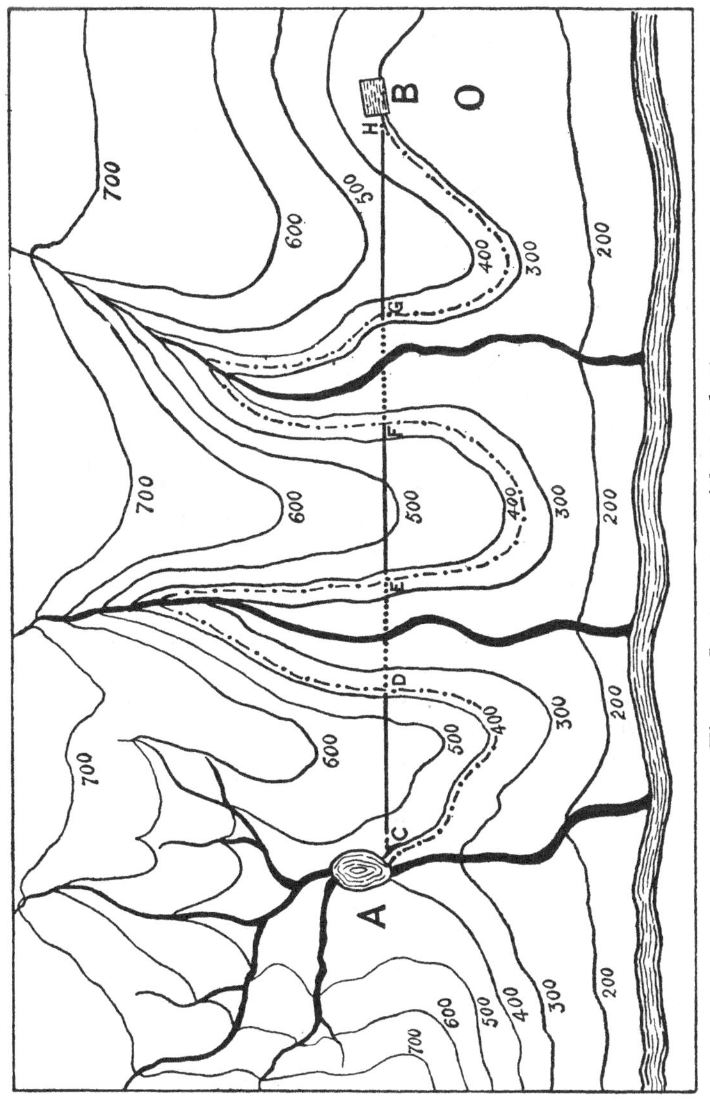

Fig. 3. Contour map with aqueduct.

uniform fall or gradient of a few inches to the mile throughout the course, though sudden falls of a few feet could be provided for by means of sluices. The channel might be roofed in so as to protect the water from contamination, but unless the whole construction were capable of sustaining great internal pressure without leaking the conduit could not be carried across a valley by going downhill and up again on the other side, so as to utilise the tendency of water to "find its own level." The open conduit had, therefore, to follow very closely the contour line of the country and to traverse up a valley till it could cross at its proper level, or to be carried round a hill unless a tunnel were bored.

Fig. 3 is a contour map of an imaginary piece of country bounded by a river which flows just below the contour of 200 feet. A large centre of population is supposed to be situated on the gentle slope near *O*. The water of the river is not suitable for domestic purposes and it lies below the level of the population. The nearest adequate supply is found in the hill basin at *A* where several streams converge to an artificial lake formed by a dam across the main stream. This is on the contour line of 400 feet. A reservoir has been constructed at *B* on the 300-feet contour just above *O*. The problem is to construct an aqueduct to carry the water from *A* to *B*. Fig. 10, p. 27, shows the course of the high and low level conduits from the springs to Jerusalem and the latter affords a good illustration of a conduit following the contour of a hilly country.

If the conduit is to be an open channel cut in the hillside the line –·–·–·– represents the course it must take with a steady fall of somewhat under 100 feet in its whole length from the contour 400 feet to the top of the reservoir on the 300-feet line. It passes round the faces of three hills and runs up and down two valleys crossing the rivulets in the valleys on two bridges. In old days these bridges would have been masonry walls with a channel at the top for the conduit and an arch below for the passage of the rivulet. The length of this aqueduct as shown in Fig. 3 is more than two and a half times the direct distance from *A* to *B*. The ancient conduit would have been lined with masonry and faced on the inside with stucco or cement to render it water-tight if the nature of the soil necessitated a lining.

If the material of the hills were moderately soft the Greeks and the Romans would probably have perforated the hills by tunnels, *CD*, *EF*, *GH*, and the Romans would probably have crossed the valleys *DE*, *FG*, by masonry arches with the water channels at the top. These masonry constructions are among the best known and most conspicuous relics of antiquity and the term "aqueduct" to many minds recalls structures of this character. On the imaginary plan these structures would be about 100 feet in height where they crossed the streams in the middle of the valleys. At the points where the tunnels open upon the valleys the modern engineer would construct small chambers from which he would lead a pipe main down the hill-side, across the stream on a bridge, or under the stream in a tunnel, and up the opposite slope to a chamber at the tunnel head made to receive it. There are a few examples known of these inverted syphons or U-tubes constructed in masonry by the Greeks, and the upper conduit which brought water from the heights near Bethlehem to Jerusalem affords a notable example of a masonry anti-syphon where it crosses the valley near Rachel's tomb. But in ancient times deep valleys could not be crossed in this way on account of the difficulty of constructing masonry to withstand great water pressure and the weakness of lead and great cost of bronze pipes, and although the Romans knew that water would find its own level they preferred the conduit raised on arches, and this held its own until cast-iron pipes were constructed. It should be noticed that as the direct line *AB* in Fig. 3 is two-fifths of the length of the open channel while the total fall available is the same, the gradient or fall per mile is two and a half times as great and consequently the water will flow more quickly and the conduit may be somewhat smaller. The sectional area might be reduced to nearly two-thirds to secure the same delivery. The student of geography should draw a vertical section through *AB* with a vertical scale of half an inch to a hundred feet. The horizontal scale may be taken to be three times that of the figure. One of the latest and most beautiful of the raised aqueducts was that of Maintenon constructed by Louis XIV for bringing the water of the Eure to Versailles. This aqueduct is 4400 feet in length and is over 200 feet in height. It comprises 726 arches arranged in three tiers, 242 in each tier (Fig. 5, p. 16).

Of the Roman aqueducts the most famous is that of the Pont du Gard across the river Gardon at Nismes constructed in the time of Augustus. Like the Maintenon aqueduct it consists of

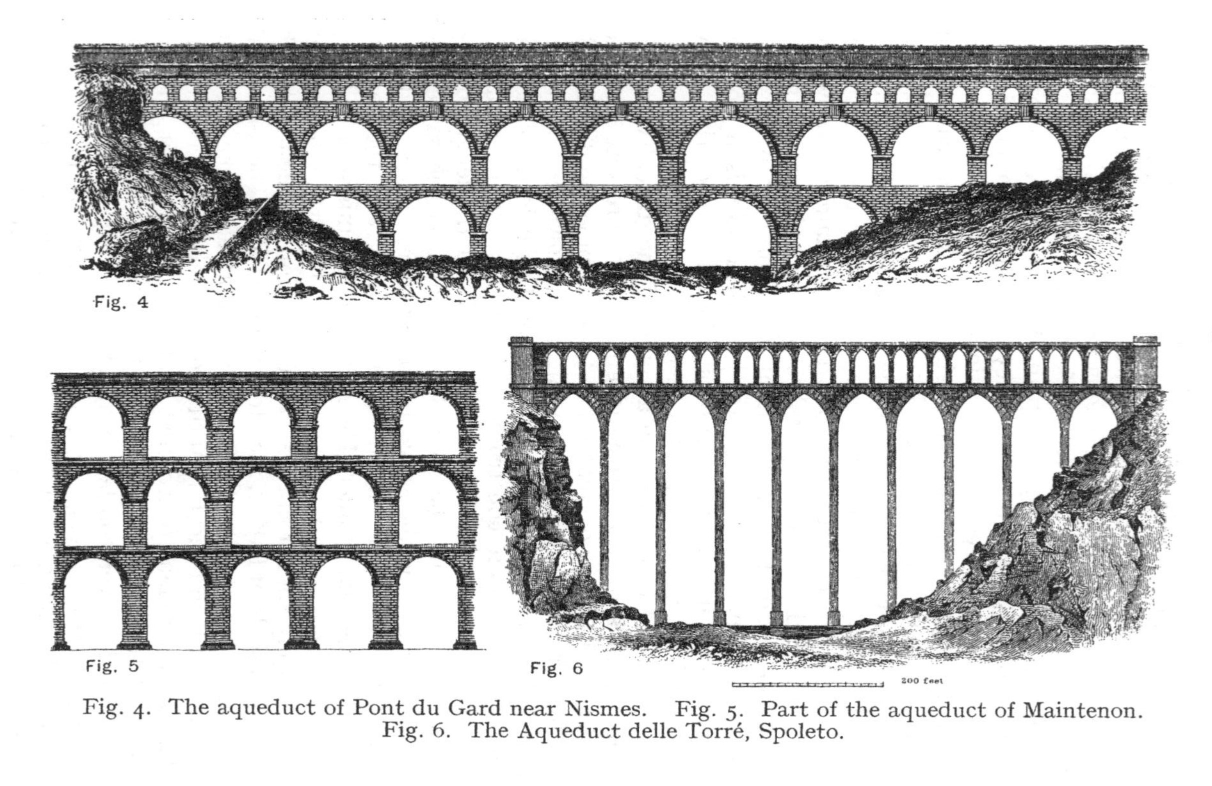

Fig. 4. The aqueduct of Pont du Gard near Nismes. Fig. 5. Part of the aqueduct of Maintenon. Fig. 6. The Aqueduct delle Torré, Spoleto.

three rows of arches carrying the conduit at a height of 180 feet from the bottom of the valley. In the lowest row there are only six arches, in the middle row 12 and in the upper 36. The aqueduct is illustrated in Fig. 4. Fig. 6 represents an aqueduct near Spoleto in Perugia, the Aqueduct delle Torré, 300 feet in height and erected in the seventh or eighth century[1]. Each arch has a span of about 66 feet. The aqueduct is now used partly as a bridge. Note its light appearance and delicately pointed arches.

In this country the finest aqueduct bridges have been constructed for the carriage of canals across river valleys. The Bridgewater canals afford excellent examples. A few miles from Manchester the canal crosses the Manchester Ship Canal, and to allow the passage of ships on the latter a section of the canal constructed in steel is raised by hydraulic lifts above the ships' masts. The section may carry with it the barge, the horse and the bargees, and when raised it is in a position very similar to the footpaths above the bascules of the Tower Bridge which connect the top of the two principal towers, and carry the suspension chains.

The Ellesmere Canal was carried across the valley of the Dee at Pont-y-Cysyllte by Telford on an aqueduct 1000 feet in length, the conduit consisting of a cast-iron channel carried on cast-iron arches supported by masonry pillars forming nineteen arches of such height that the surface of the water in the canal is 126 feet 8 inches above the water in the Dee. Many other canal aqueducts might be mentioned, but they do not belong to the subject of water supply and there is in general no flow of water along them, the conduits being constructed on a dead level.

For the supply of the city of Rome no less than eleven aqueducts were constructed by the Censors during the republic and by the emperors. Some of these were carried on long aqueduct bridges. The best known is the Porta Maggiore constructed by the Emperor Claudius which carried two conduits built in masonry one above the other bringing water from separate sources, the Aqua Claudia and the Anio Novum. The source of the first supply was two springs in the Sabine Hills 35 miles from Rome, though the length of the conduit, mainly underground, was 45 miles. The second source was the river Anio,

[1] For these three illustrations the author is indebted to the *Encyclopaedia Britannica*.

62 miles from Rome. The two conduits were united on the Porta Maggiore about six miles from the city. The Romans appear to have known the value of storage reservoirs for the purpose of purifying the water supply.

In a modern aqueduct when a tunnel is cut through a hard rock which is watertight and free from fissures and which is not partially soluble in the water no lining is necessary. The tunnel is generally of the shape shown in Fig. 7. When the rock is porous it is lined with masonry which is rendered on the inside in cement and the space between the masonry and the rock is filled up with rubble masonry. When the water is required for domestic supply open cuttings are now invariably

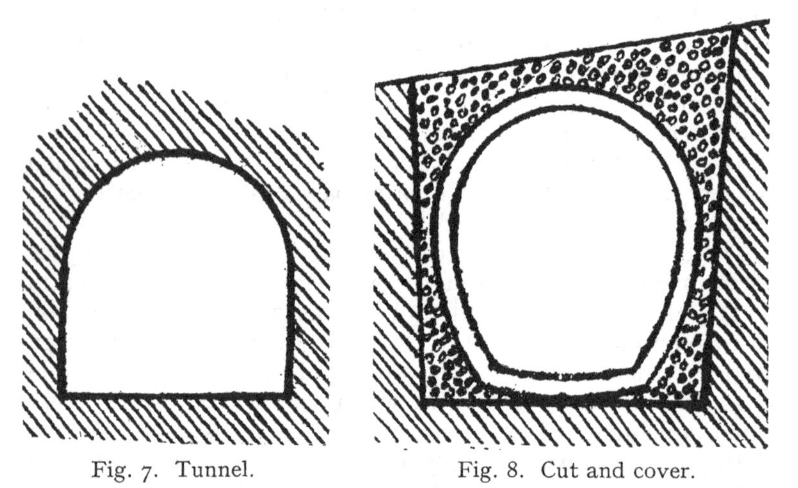

Fig. 7. Tunnel.

Fig. 8. Cut and cover.

roofed over to prevent contamination of the water. Earth is usually filled in on the top of the roof so as to conceal the scar on the hillside, and the construction is known as "cut and cover." When the rock is porous it is lined with masonry in the form of a horseshoe as in the case of a tunnel through porous rock (Fig. 8). When a valley is crossed by a high-level bridge the conduit may be a trough built of steel plates riveted together and carried on piers of brickwork or masonry or the water may be carried in cast-iron pipes laid in a channel formed on the top of an arched bridge.

When the water is taken across a valley by an inverted syphon, cast-iron socketed pipes with lead joints are usually employed, the thickness of the pipes depending on their diameter and the pressure of water they have to sustain, but sometimes the pipes are built up of steel plates like the body of a Lancashire boiler. For a bursting pressure steel is much stronger and more reliable than cast-iron and a much less weight of metal can be employed, while the pipes can be made of greater diameter. These riveted steel pipes are largely employed in Switzerland for carrying water from mountain sources for electric generating stations, especially for the electrification of railways. The generating station for the Simplon Tunnel at Brigue is supplied with water at high pressure carried across the Rhone in a riveted steel tube on a steel bridge constructed for the sole purpose of carrying the conduit. The traveller to Lauterbrunnen can scarcely fail to notice the steel pipe which brings the water to the generating station of the Jungfrau Railway by the side of the Lutschine.

In some parts of the western States of America where timber is plentiful, but iron and steel involve very costly carriage, high-pressure conduits are constructed of thick planks of timber fitted together like the staves of a barrel and hooped with steel, but the planks are so arranged as to break joint when they abut end to end. In this way the conduit is built in one continuous length of barrel, and it can be constructed to withstand very great pressure so as to carry water across a valley in an anti-syphon or to bring it down a mountain side to drive a high-pressure turbine. The end joints between the staves are made with an iron tongue to render them watertight. The hoops are tightened by bolts and nuts.

The city of Denver is supplied through wooden pipes constructed in this way, one 34 inches and the other 30 inches in diameter and conveying together 30,000,000 gallons a day. These pipes are 16½ miles in length and sustain a pressure of 185 feet of water. Salt Lake City is supplied through a wooden pipe six feet in diameter and sustaining a pressure in some parts of 117 feet of water.

The Water Supply of New York

The principal water supply of New York City is derived from the hill country on the border of the States of New York and Connecticut, between that border and the River Hudson. The

Croton Creek joins the Hudson a little more than 35 miles from New York. This creek receives water from a hill basin with an area of more than 300 square miles below the Catskill Mountains. The collecting reservoir was made by building a dam across the creek. The lake so formed has an area of 400 acres and a capacity of 500,000,000 gallons. The first aqueduct, 38¼ miles in length, was constructed between 1837 and 1842. It consisted mainly of tunnel lined with masonry and brickwork so as to form a horseshoe conduit 8 feet 5 inches in height with a maximum width of 7 feet 8 inches. The water was carried over the Harlem River, 33 miles from the lake, in iron pipes laid on a masonry bridge 1460 feet in length and 114 feet above the river. This aqueduct was capable of discharging 95,000,000 gallons a day. In 1880–1885 a smaller aqueduct, pipe line, was constructed from the Bronx River, at a distance of about 27 miles from New York. This supplied about 28,000,000 gallons a day, but in 1885 the New Croton Aqueduct was commenced. This was capable of conveying 302 million U.S. gallons a day. The U.S. gallon is ·833, or nearly five-sixths, of the Imperial gallon. It is 231 instead of 277·274 cubic inches. The difference in the U.S. and Imperial gallon has led to considerable trouble in connection with contracts for oil.

Like the first Croton Aqueduct the New Aqueduct consists mainly of tunnel. Through 23·7 miles of tunnel the water flows with a free surface and this part of the tunnel is of horseshoe form 13 feet 7 inches in height and width, except where the rock is very friable and the circular form of brick lining has been adopted. A little over seven miles of tunnel carries the water under pressure and is circular in form and 12 feet 3 inches in diameter. Only nine furlongs consist of "cut and cover," that is, open trench lined with a covered masonry or brick conduit and covered over with soil. There are nearly 2½ miles of pipe-main consisting of eight rows of pipes 48 inches in diameter. The tunnels and the cut and cover are lined with brickwork, set in cement mortar, plastered on the inside with cement and sand and rendered with neat cement to make the conduit water-tight.

Like the old aqueduct the New Croton Aqueduct had to cross the Harlem River about four miles from the reservoir in the Central Park at New York, but it was not carried across on a bridge. Of late, engineers have preferred, where possible, to pass under streams rather than cross them on bridges. For passenger traffic the several tunnels under the Thames are examples of this tendency, and the New Croton Aqueduct was no exception. The general gradient of the conduit is ·7 foot per mile (8·4 inches), but, about four miles before reaching the Harlem River crossing, the tunnel slopes downwards at a gradient of 10 per cent. (528 feet to the mile) and descends on this slope for nearly 120 feet. It then continues as a circular tunnel, 12 feet 3 inches in diameter, with a gradient of ·7 foot per mile until very near the bed of the river where it again slopes rapidly downwards. From the bottom of this incline a vertical shaft is sunk to a further depth of 174 feet, a similar shaft being sunk 1300 feet away on the further side of the river. These two shafts are connected by a tunnel 10 feet in diameter at a depth of 300 feet below high-water level in the river, but the shaft on the city side of the river rises vertically to a height of 268 feet. From the top of this shaft the tunnel rises by a gentle gradient until it approaches the Manhattan Valley when it rises rapidly to a tower and gatehouse from which the cast-iron pipe-mains start and cross the valley to the Central Park reservoir.

In order to provide sufficient storage a new dam was built across the Croton stream below the former dam so as to include some additional collecting ground, and was carried to a much greater height than the former dam. The total height of this dam from its lowest foundation is 294 feet and the water-level is 149 feet above the river beds. The quantity of water thus impounded is 26,000,000,000 gallons. In addition to this, about sixteen other storage reservoirs were constructed by building dams across the tributary streams in higher parts of the valley, the storage provided representing over 70,000,000,000 gallons.

The tunnel from the lake to New York was constructed by sinking thirty-six shafts at intervals averaging a little over a mile and working in both directions from the bottom of each shaft. In the lower part of the course above the deep tunnel some of the openings of the shafts are below the hydraulic gradient, that is, the level to which the water will rise while the steady flow of 302 million gallons a day is maintained in the tunnel. Roughly, the level is lower than the water-level at the inlet from the Croton reservoir by 8·4 inches per mile of conduit. These shafts have to be closed with double watertight doors to prevent the water escaping.

The Ancient Water Supply of Jerusalem

The City of Jerusalem lies on a promontory which stretches south from a slight ridge or water parting a little to the north of the city wall at a height of about 2550 feet above the Mediterranean and is bounded on the east by the valley of the Kidron, which starts from the same ridge, and on the west by the valley of the Son of Hinnom which has its origin near the same ridge and, passing the west side of the promontory, curves to the east and joins the valley of the Kidron to the south of the south-east corner of the city wall about 2000 feet above the sea. Just outside the Damascus gate, which is near the middle of the north wall, a third valley, the Tyropoeon of Josephus, starts, and sloping downwards to the south through the centre of the city, opens out upon the other two valleys opposite their point of union and near the Lower Pool of Siloam. The portion of the promontory between the Tyropoeon and the valley of the Kidron forms the East Hill, and it is upon this that the temple area is situate, while that between the Tyropoeon and the Valley of Hinnom is known as the West Hill. Principal Sir George Adam Smith says: "The two may be roughly likened to a thumb and forefinger pointing south, the latter somewhat curved. The hand in which they merge to the north is the plateau."

The whole length of the promontory is about a mile and a quarter. The base of the north wall of the city at its highest point is nearly 2500 feet above the Mediterranean. The lowest point of the south wall is at the level of about 2300 feet as it crosses the Tyropoeon, so that there is a fall of nearly 200 feet within the city walls.

South of the city the two hills slope down rapidly to the point of junction of the three valleys, about 500 feet below the highest point of the city. The central valley, or Tyropoeon, is almost filled up with the building *débris* of the city, while the valley of the Kidron is similarly choked to a depth varying from 10 to 50 feet. On the east of the Kidron valley are the Mount of Olives, 2693 feet, and the Mount of Offence. To the south of the valley of Hinnom is the Hill of Evil Counsel, 2550 feet. The steepness of the sides of the promontory as they slope to the valleys made it very difficult for Jerusalem to be attacked except from the north where the ground slopes gently to the city from the plateau. This, no doubt, determined the selection of the site by the Jebusites as a stronghold. In addition to inaccessibility a city which is to stand a siege must have an efficient water supply accessible from within its walls and not easily cut off by an enemy outside.

The promontory consists entirely of limestone and chalk belonging to the periods of the lower, middle and upper chalk of Britain. Over a considerable portion of the city the upper stratum consists of a hard limestone or marble used for building stones. Under this lies a stratum of very soft limestone or chalk which has a thickness of about 35 feet and underlies the whole city. Beneath this is a bed of very hard limestone, impervious to water except where it is fissured. This limestone is exposed in the valley of Hinnom and to a less extent in the bed of the Kidron. All the strata dip to the east or south-east at an angle of about 10°. This causes any water which accumulates in the soil of the promontory to flow eastwards towards the Kidron valley.

The West Hill is somewhat higher than the East Hill, the greater part of the surface standing more than 2500 feet above the Mediterranean. The citadel on the West Hill stands at 2542 feet, or just 102 feet above the temple area, on the East Hill. The principal water parting of Palestine runs nearly north and south a little to the west of Jerusalem and reaches a height of more than 3000 feet about 10 miles south-south-west of the city. Here, as we shall see later on, is afforded an opportunity of collecting water from the hills to be conveyed by aqueducts to Jerusalem, but in the early days of the city such aqueducts had not been formed. The Kidron could not be relied upon for a water supply, for it flowed only after heavy rains and chiefly in the winter. During the greater part of the year the rain which fell on the plateau north of Jerusalem sank into the chalk just as much of the rain falling on the chalk hills of Surrey and Hertfordshire finds its way underground into the London basin.

As far as is known the only perennial water supply available in the earliest days of the city, without going some distance down the valley, was the Virgin's Spring, the Gihon of the Old Testament. This spring was, and is, situated on the side of the East Hill a little above the Kidron, some 350 yards south of the temple area and just below the site of the Jebusite fortress of Sion. The spring, or fountain, is situated in a cave above the natural floor of the Kidron valley but below the level of the rubbish with which the valley is covered and by which the original bed is concealed. The cave is approached from the present surface of the valley by two flights of sixteen and fourteen steps. The basin of the spring forms a tank about 50 feet in length and 8 feet in breadth. The water enters by an opening in the rocky floor near the middle of the tank. Formerly, the overflow was into the Kidron valley, but the filling-up of the valley with the *débris* of the hills and the building material of the city has rendered this impossible and the water flows away by the tunnel of which we shall speak presently into the Upper Pool of Siloam in the Tyropoeon and within the walls of the city. At one time the overflow into the Kidron valley appears to have been carried by an aqueduct to the Lower Pool of Siloam near the point at which the three valleys join. The flow of water into the basin is not continuous. During a rainy season a rush of water takes place three, four or five times in the day. In very dry seasons only once a day or even less. In the early days it was believed that the water was swallowed by a dragon, one of the type that produced earthquakes and disturbances at sea, but when the dragon dropped asleep the water rushed past him and filled the cistern. The more modern view is that there is concealed somewhere within the hill and at a higher level than the spring a natural cavity in which the water, filtering through the soft limestone, collects and that this is connected with the spring by a natural syphon which comes into action as soon as the water in the cavern rises above the top of the syphon, which then continues to discharge the water until its level is reduced below that of the syphon mouth, the action being precisely similar to that of an automatic flushing tank. With such an arrangement as is indicated in the sketch (Fig. 9), as soon as the water collected in the cavity *X* reached the level *AB* it would flow into the basin *Y* until the level in *X* fell to *CD*. The flow would then cease until the water rose again to the level *AB* so as to overflow the hump *Z*, which is the modern substitute for the dragon.

From the cave there is a very ancient tunnel 50 feet long leading into the hill; at the end of this tunnel a shaft rises vertically to a height of 44 feet and from the top of the shaft a gallery at first runs horizontally and then slopes steeply upwards through the rock to the surface of the hill 170 feet above the spring. It is most probable that the opening of this gallery was within the Jebusite fortress of Sion and that the gallery and shaft were made to enable water to be raised by buckets and brought into the fortress for the use of the garrison during times of siege. It is not improbable that this gallery served as the approach by which David and his men took Jerusalem as recorded in II Sam. v. 6–9: "And the king and his men went to Jerusalem against the Jebusites....David took the stronghold of Zion; the same is the City of David. And David said on that day, Whosoever smiteth the Jebusites, let him get up to the watercourse...and David dwelt in the stronghold and called it the City of David." (Revised Version.)

Later on a tunnel was cut right through the eastern hill from the Virgin's Spring to the Upper Pool of Siloam. In a straight

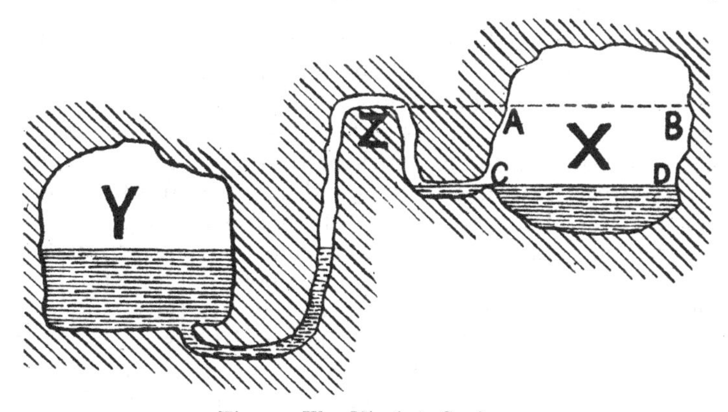

Fig. 9. The Virgin's Spring.

line the distance is about 1100 feet, but the tunnel, which varies in height from 11 feet to less than two, now that the floor is in some places deeply covered with *débris*, winds considerably and has a total length of 1750 feet. This tunnel is the present outflow of the spring. The tunnel, like the Simplon tunnel, was started from both ends and indicates by its course that it was not easy for the excavators to meet. An inscription in ancient pre-Exilic characters, discovered on the wall in 1880, records the meeting of the two parties: "When yet there were three cubits to be bored heard was the voice of each calling to his fellow; for there was a fissure...the hewers struck, each to meet his fellow, pick against pick; then went the waters from the issue to the pool for two hundred and a thousand cubits, and a hundred cubits was the height of the rock above the hewers."

This tunnel appears to have been the work of Hezekiah. In II Kings xx. 20 Hezekiah is reported to have "made the pool and the conduit, and brought water into the city"[1]. There was a double object in this. In a time of siege the Kidron valley might be occupied by the enemy. It was essential that the people of Jerusalem should get water without going outside the walls and almost equally important that the enemy should not too readily find a supply. So Hezekiah "sealed all the springs and the Nahal flowing through the midst of the land, saying, Why should the Kings of Assyria come and find much water?" (II Chron. xxxii. 4). It was "when Hezekiah saw that Sennacherib was come, and that he was purposed to fight against Jerusalem, he took counsel with his princes and his mighty men to stop the waters of the fountains which were without the city" (II Chron. xxxii. 2, 3). The *Nahal* or "brook" was the intermittent stream which flowed down the Kidron valley.

The well of Job is a well 125 feet deep sunk in the porous limestone about three furlongs south of the Virgin's Spring. It is said that water is always to be found in this well into which it filters as water filters into chalk wells in this country.

Cisterns for collecting rain water from the roofs of buildings and the surface of the ground were made in the limestone rock at all times, but of these the most famous are those under the temple area. No less than thirty-seven cisterns or reservoirs have been examined there. Of these, "the Great Sea" which is 60 feet in depth has a capacity of about two million gallons.

As in other cities with the increase of population it became necessary to search for water farther afield. The main supply within a reasonable distance was to be found south-south-west of Jerusalem on the water parting of Palestine beyond Bethlehem where the hills rise more than 3000 feet above sea-level. Water was to be found in other directions among the hills, but the levels were not generally suitable for the supply of the city by gravitation. From these hills a steady fall could be obtained from the springs to the higher parts of the city.

The three reservoirs known as the Pools of Solomon, about two miles south-west of Bethlehem, were probably constructed by Herod the Great. They are about 2616 feet above sea-level.

[1] Compare II Chron. xxxii. 30.

Five springs in the immediate neighbourhood are available to supply these reservoirs which are about six and a half miles from Jerusalem. Farther south and about 12 miles from Jerusalem is another group of springs with a reservoir at a height of 2740 feet. These groups of springs were connected by conduits with two aqueducts leading the waters to the city (Fig. 10). The water from the most distant springs which, if not intercepted, would flow in a river to the Dead Sea, is carried by an aqueduct which winds along the sides of the several valleys to the middle pool of the three Pools of Solomon. The length of this aqueduct, shown by a full line in Fig. 10, is about 25 miles to Solomon's pools though the direct distance from its upper extremity is only five miles. Its course resembles the original route of the New River or of Drake's conduit at Plymouth, but the winding is more exaggerated on account of the number of valleys to be passed. The low-level aqueduct continues from the lowest of Solomon's pools and traverses the sides of the valleys, except that it passes through the Hill of Bethlehem by a tunnel; after a second tunnel it traverses the western slope of the valley of Hinnom, crossing the valley on arches; it then turns back on the eastern slope and passing round the southern end of the western hill enters the city and finally crosses the Tyropoeon to supply the cisterns under the temple area. This aqueduct appears to have been constructed by Herod. Its total length is over 40 miles, as against a direct distance of 12 miles.

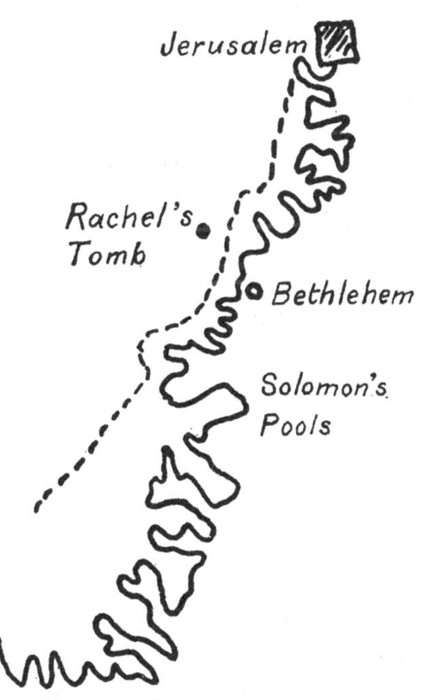

Fig. 10. High and low-level aqueducts.

The high-level aqueduct, shown in Fig. 10 by a broken line, brought water from another group of springs about two miles south-west of Solomon's pools through tunnels, and passing the pools about 150 feet above the low-level aqueduct and to the west of Bethlehem crossed the valley in which is situate the tomb of Rachel by an inverted syphon built of limestone blocks each perforated by a circular orifice 15 inches in diameter. The blocks were imbedded in rubble masonry. There are other examples in the East of inverted stone syphons carrying waters across valleys and anticipating the modern iron pipes. This construction and the free use of tunnels enabled the high-level aqueduct to reach the city in a much more direct course than the other. The date of this aqueduct is uncertain, but it seems probable that it was constructed somewhat earlier than the low-level aqueduct of Herod. It is now disused, but the low-level aqueduct still carries water to Jerusalem.

The two real aqueducts shown in Fig. 10 may be compared with the imaginary ducts in Fig. 3, p. 13, as indicating the effect of tunnel and anti-syphon as compared with an open channel following the contours.

This brief account of the ancient water supply of Jerusalem has been condensed from the work of Principal Sir George Adam Smith.

Mountain Sources of Water Supply

The mountains of Wales and Cumberland consist very largely of slate rock covered with a very thin layer of peaty soil. Here and there in the valleys are lakes, the basins of which were scooped out ages ago by glaciers when the Welsh mountains were covered with snow. The lakes are fed by streams from the mountains and the overflow forms a river running down the lower part of the valley. As a rule, there are mountains on each side of the lake, a valley behind the lake which descends from a pass, which is a depression in the ridge forming the water parting, and a descending valley in front of the lake down which the overflow from the lake descends in a river.

The Water Supply of Manchester

Forty years ago this was the condition of Thirlmere in Cumberland, a lake lying at the foot of Helvellyn and receiving water from about 11,000 acres. The Manchester Corporation built a dam across the valley where the water flowed out from the lake, carrying the dam to a height of 104½ feet above its foundations for a length of 857 feet. The dam can keep back the water so as to make the lake 50 feet deeper than before. This causes the water to rise up the sloping sides of the mountains and up the valley at the upper end of the lake so as to increase the length of the lake to 3½ miles and to increase the area of the water surface from its natural area of 330 acres to 793 acres. Up to the end of 1915 the water had been allowed to rise only 35 feet and the surface of the lake had been increased to 690 acres. Raising the level of the lake 50 feet as at present gives a quantity of water over and above the original contents of the lake amounting to 8,135,000,000 gallons from which Manchester may draw without lowering the water below the original level. The water from this reservoir is carried to Manchester partly in tunnels, partly in cement conduits about 7 feet square and partly in iron pipes. When it has to cross a valley it is carried in iron pipes forming a great U-tube down one side and up the other where it joins the cement conduit again. The height of the surface of the lake above Manchester, when raised 50 feet, is about 584 feet; this gives ample fall to carry the water along the 96 miles of conduit, of which 14½ miles are tunnel, 7 feet in diameter, 45 miles rows of iron pipe 36 to 48 inches in diameter, and 36¾ miles cement conduit, with a fall of 20 inches to the mile. This is Manchester's New River. There were in 1916 two conduits each capable of carrying 10,000,000 gallons a day and a third was in course of construction. Eventually there will be five lines side by side.

By the side of the lake, a little higher than the water can ever rise, is built the well-house. In the middle of this house is the well which is a cylindrical shaft about 37½ feet in diameter and 65 feet deep. Near the bottom the water from the lake enters the side of the well, but before it can flow out on the other side it has to pass through screens of copper-wire gauze of which seven are arranged in frames to form seven sides of an octagon. When the water has got inside this octagon through the gauze it flows out by the eighth side into the tunnel which carries the water under the ridge of Dunmail Raise which forms the water parting between Thirlmere and Grasmere. Hence the water flows to two reservoirs having a joint capacity of 40,000,000 gallons at Prestwich near Manchester. The rainfall at Thirlmere is nearly the greatest in England.

About 200 feet above the well-house on the side of Helvellyn, a mountain stream fills a small reservoir from which a pipe brings the water at high pressure into the well-house. This water is used to drive a hydraulic crane by which the gauze screens in their heavy frames are lowered into the well and raised when they require cleaning. A jet of the same high-pressure water directed from a hose is used for scouring the wire gauze screens and removing all the *débris* which has been collected on the gauze.

Nearly twenty-five years before the Thirlmere works were commenced the Manchester Corporation constructed five reservoirs in Longdendale, on the river Etherow, about 16 miles east of Manchester. These reservoirs have an area of about 859 acres in a region where the rainfall amounts to about 50 inches a year. They contain about 6,000,000,000 gallons and the supply is equivalent to 25,000,000 gallons a day. This source of water is still employed to supplement the Thirlmere supply. The water, like that from Thirlmere, is very soft and well adapted for dyeing and bleaching. It is said that in Manchester the change from hard water to soft effected a saving in soap and soda of between £80,000 and £100,000 a year. The Thirlmere supply was first delivered in Manchester in 1894.

The Water Supply of Glasgow

The first scheme carried out in Britain for the water supply of a great city from a distant upland source was that for the supply of Glasgow from Loch Katrine. The work was commenced in 1855 and completed in 1860, at a cost of £668,000, exclusive of the purchase of land. The surface of the water in Loch Katrine was raised by 4 feet by means of a dam, arrangements were made to draw off the water if necessary to a depth of 7 feet below the new surface, that is, from a level of 360 feet above ordnance datum. This afforded a storage capacity of 5,687,500,000 gallons. The levels of Loch Venachar and Loch Drunkie were also raised in order that they might serve as compensation reservoirs. The service reservoir at Mugdock from which the Glasgow supply is immediately drawn is 8 miles from the city and the length of the aqueduct from Loch Katrine to the reservoir is 25¾ miles. This aqueduct is partly tunnel, 13 miles, and partly "cut and cover," 9 miles, 8 feet in height and in width. The iron pipes across the valleys represent $3\frac{3}{4}$ miles. It was calculated to convey 50,000,000 gallons a day with a gradient of 10 inches to the mile. Narrow gorges are crossed on aqueduct bridges and broader valleys by anti-syphon pipes, two of 4 feet diameter and one of 3 feet. The very wild character of the country near Loch Katrine, with no roads and no building material, greatly increased the cost and difficulty of the work of construction. The first ten miles consisted of primary rocks of mica schist and clay slate which in those days were blasted with gunpowder. The deeper portions of the mountain gorges are crossed by wrought-iron conduits 8 feet by $6\frac{1}{2}$ feet carried on piers, and the shallower portions by cast-iron conduits supported on a masonry wall. At the bottom of the Endrick valley the cast-iron pipes forming the anti-syphon are exposed to a pressure of 235 feet of water. As usual, the longer tunnels were constructed with the help of shafts sunk from the surface, each providing two new faces from which the work of tunnelling could proceed. There were forty-four of these shafts and in all eighty tunnels, some very short. The longest was a mile and a half in length. There were twenty-five important bridges of iron and masonry.

The service reservoir at Mugdock has an area of 60 acres and a storage capacity of 548,000,000 gallons. This allows repairs to be carried out on the aqueduct without interfering with the water supply of the city. The surface of the water when the reservoir is full is 317 feet above ordnance datum, that is, 50 feet below the highest level of the water in Loch Katrine. The water is strained through copper gauze in a well at the Mugdock reservoir just as the Manchester water is strained in the well at Thirlmere, described on p. 29. The water, which is collected on primary rocks, is very pure and soft like that of Thirlmere.

Watersheds and Water Partings

Reference has already been made, p. 29, to the water parting at Dunmail Raise, between Thirlmere and Grasmere, under which the Manchester water supply is carried in a tunnel. The area from which a river collects its water is often called the watershed of that river. When we are speaking of the collecting area for water supply, which is usually the upper portion of the watershed, it is commonly called the gathering ground. The watershed of a river is generally bounded by the highest ground in the neighbourhood, and this is called the water parting, or, in America, the divide. There are, however, many examples of rivers passing through gaps in ranges of high hills, which may be higher than the water partings. Such examples are found in the North and South Downs and in the Isle of Wight, where the Medina and the eastern and western Yar pass through gaps in the range of chalk downs which extends from the Culver Cliff to the Needles. On the farther side of a water parting the ground slopes in the opposite direction and the water generally drains away into another river so that the water partings form the lines of separation between the river basins of the country. Commonly the term watershed is applied to the water parting, but it is better to employ it solely for the collecting ground of a river. In low-lying countries the water partings are not very conspicuous, but in mountain regions they are formed by mountain peaks and ridges the lowest portions of which form the passes from one river valley to another. Where the sides are not too steep large fields of snow accumulate on the slopes of mountains above the snow-line and slowly descend into the valleys between them, sometimes, especially in the spring, coming down with a rush as an avalanche. This snow accumulates in the high valleys and the pressure converts it into solid ice so that much of the higher portions of a great mountain range consists of fields of ice through which rise the lofty peaks and ridges which form the water, or in this case the snow, partings. The ice as it is constantly fed by fresh snow flows very slowly down the valleys past green slopes, or Alps, and sometimes growing timber, as glaciers or ice rivers. These generally become narrower as they descend. Winding with the valleys the ice cracks first on one side and then on the other, and as the surface melts in the warmer air and sunshine the water flows down these cracks or crevasses to the bed of the glacier where it flows through an ice tunnel made by itself until it emerges from an ice cave at the foot of the glacier, and this is how mountain rivers are born. The sketch-map in Fig. 11 indicates these ice fields in the Upper Valais between the Jungfrau group and the Rhone Glacier, through which rise some of the highest and sharpest peaks of the Oberland. It also indicates the position of a few of the principal glaciers. The Aletsch is the largest in Switzerland, descending from the south-easterly slopes of the

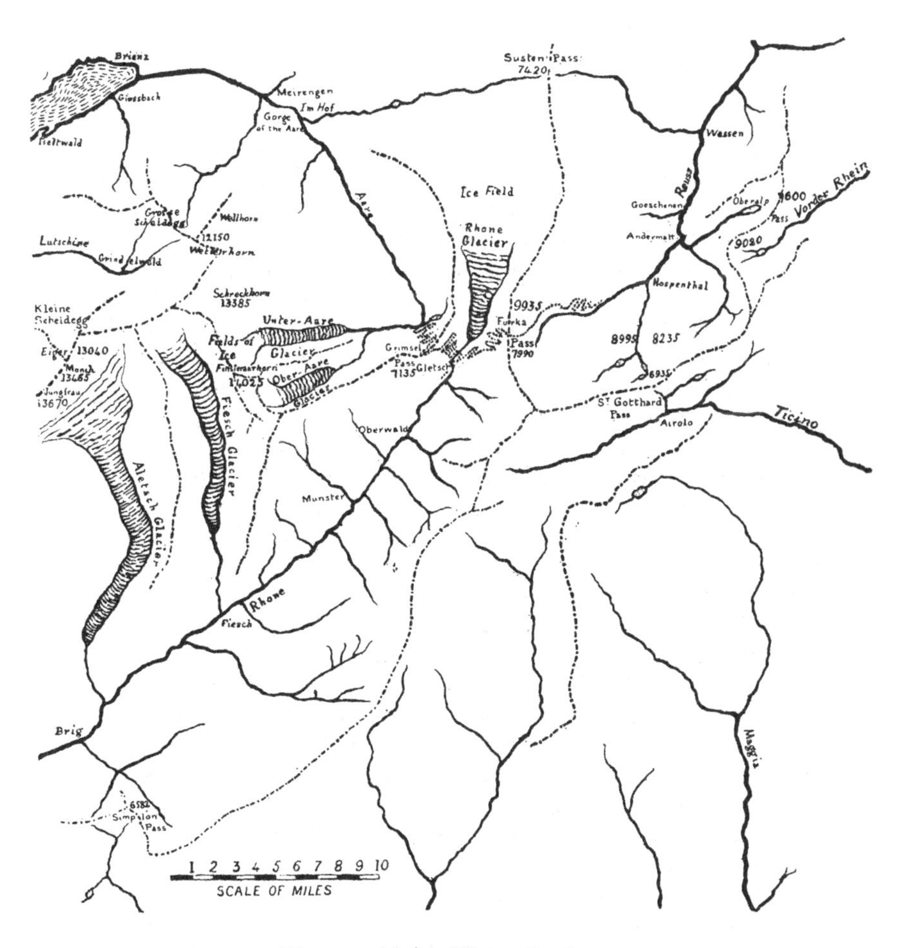

Fig. 11. Alpine Water Partings.

Jungfrau and Monch. Some of the chief water partings are shown by dot-and-dash lines and the places where these water partings are crossed by well-known passes are indicated in a few cases. As the glaciers move slowly down the valleys, 50 feet or more in a year, they carry, frozen into their mass, some of the rocks and stones which fall upon them from the mountain-sides and these cause the bottom of the glacier to act as a file and gradually wear away the rock below, deepening and widening the valleys and charging the glacier streams with the fine dust abraded from the valley. At one time the Welsh and Cumberland mountains were covered with a sheet of ice and glaciers flowed down the valleys. Some of the lake basins appear to have been scooped out there, as in the Alps, by the glaciers which receded or disappeared altogether with the advent of a warmer climate. Most of the Swiss glaciers are receding now. The Rhone Glacier has receded by more than a mile during the last century. The Gletsch Hotel was built close to its foot. It is now more than a mile away.

The sketch-map shows in its central portion the great water parting of Central Europe, the mass of the St Gotthard. This, the oldest part of the Alps, covers about 160 square miles and comprises several peaks besides the Sasso di San Gottardo. Fig. 12[1] is a picture of the Rhone Glacier with the Rhone flowing from its end. To the right are seen the zigzags of the road from Gletsch to the Furka Pass. The road to the Grimsel lies to the left behind the hotel and comprises similar zigzags. These roads, leading up from the Rhone valley over the two passes, are shown on the map, Fig. 11, by simple dotted lines. They are not traced beyond the steep slopes, because in the valleys they follow very closely the lines of the rivers. We can conveniently make our examination of the map by starting from Gletsch. Crossing the Grimsel Pass at a height of 7135 feet, and descending the zigzags, we come upon one of the sources of the Aare. The principal sources are the Ober-Aare and Unter-Aare glaciers descending from the slopes of the Oberaarhorn and the Finsteraarhorn. The Aare flows past the Handegg Falls, through a marvellous gorge between Im Hof and Meiringen, and thence in a canalised bed to Lake Brienz, through Interlaken and Lake Thun to Berne

[1] This and many other Alpine views beautifully coloured and illustrating mountains, rivers and glaciers can be obtained from the Photochrom Company, Tunbridge Wells.

Fig. 12. The Rhone Glacier.

and thence northwards to join the Rhine between Schaffhausen and Basle, picking up the Reuss about ten miles from the Rhine.

If at Gletsch we turn to the right and cross the Furka at a height of 7990 feet we have just below us the source of the Reuss flowing on to Hospenthal, where it is joined by a stream from the St Gotthard Pass, and thence to Andermatt where it receives the overflow of the Ober-Alp Lake. From Andermatt it passes through the Schöllenen Gorge and on past Altdorf to enter the Lake of Lucerne near Flüelen. It leaves the lake at Lucerne and joins the Aare as above mentioned.

If we leave the Reuss at Andermatt and ascend the zigzags of the Ober-Alp, passing round the end of the Ober-Alp Lake and turning south-east we cross the Ober-Alp Pass at a height of 6720 feet and soon find ourselves close to the Vorder Rhein, which flows through Lake Toma. The Rhine flows first to the east and then turning north passes through Lake Constance and Schaffhausen and Basle, picking up the Aare and Reuss, and thence through Germany and Holland to the North Sea.

If we retrace our steps to Hospenthal and turn south we cross the St Gotthard Pass at a height of 6935 feet and descend to Airolo in the valley of the Ticino. This river, which rises a few miles west of Airolo, receives the drainage from the lakes on the south side of the St Gotthard Pass, and then flows south-east to Bellinzona and thence to Lake Maggiore, whence it is discharged by the Po into the Adriatic. The other rivers shown on the map as flowing to the south also make their way to Lake Maggiore.

Returning now to Gletsch and the Rhone Glacier we are facing the source of the Rhone as the stream emerges from the glacier cave. Flowing through Gletsch and Münster the Rhone receives at Fiesch the stream from the Fiesch Glacier and a little lower down, near Brig, that from the Great Aletsch Glacier. A very little beyond the western limit of the map the Rhone receives the Visp with the water from the glaciers of the Matterhorn, Monte Rosa, the Dom and the other mountains of that group, and flowing through the Lake of Geneva it finds its way to the Gulf of Lyon.

The St Gotthard mass thus forms the parting between rivers which flow to the North Sea, the Adriatic and the Gulf of Lyon. As stated above, some of the other principal water partings are indicated on the map by dots and dashes, while the figures correspond to heights in feet. It should be specially noticed how close together in the middle of the map are the sources of the Ticino, the Reuss and the Rhone and the bend of the Aare at the Grimsel. We shall have occasion to notice water partings in the south of England, but we have nothing comparable with the great parting of the St Gotthard.

The Liverpool Water Supply and Masonry Dams

In going to the Welsh mountains for their water supply the corporation of Liverpool did not choose an existing lake but set to work to make a lake for themselves. They selected a mountain valley down which a river was running. This was the Vyrnwy, a tributary of the Severn, which received its water from the surrounding mountains. These are of hard slaty rock from which the water flows very quickly without soaking in or dissolving the rock, so that the water in the river is almost pure rain water. There is reason to believe that the broad part of the valley which was chosen for the Liverpool reservoir was once a lake formed, like the Swiss lakes, by a glacier when the Welsh mountains were covered with snow. This lake had been filled up with soil through which the river flowed, but stones belonging to the old glacier moraine were found below the surface. In the midst of the valley were a small village, Llanwddyn, and several farms, the buildings of which had to be removed. The water in a lake basin formed by a glacier is held back by a bar of rock across the lower end of the lake and the outflow of the lake is over this bar. The engineers found the bar 50 or 60 feet below the surface by boring through the soft soil above it. A trench was dug down to this bar 120 feet wide and extending from one side to the other of the valley, a distance of 1172 feet. In this trench was laid the foundation of the great stone dam which was built to a height of 152 feet above the foundations with a carriage road along the top, but the water overflows 18 feet below the road or 134 feet above the foundations. The water is therefore between 70 and 80 feet deep where it presses on the dam. When the dam had been completed, the village removed and the bottom of the valley cleaned up, everything was ready for the lake to be filled. It was then only necessary to build up the opening in the dam which had been left for the river to flow through and the water was bound to accumulate in the valley

Fig. 13. The site of Lake Vyrnwy.

Fig. 14. Lake Vyrnwy.

until it reached the overflow openings near the top of the dam at a height of nearly 80 feet above the bottom of the lake. The reservoir is known as Lake Vyrnwy, after the name of the river. It is 821 feet above the sea, has a length of nearly 5 miles and a surface area of 1121 acres. It contains 12,000,000,000 gallons of water.

Fig. 13 shows the mountain valley in 1888 before the water was allowed to accumulate but after the dam had been built. Fig. 14 shows the same valley in December 1889 when the reservoir was full.

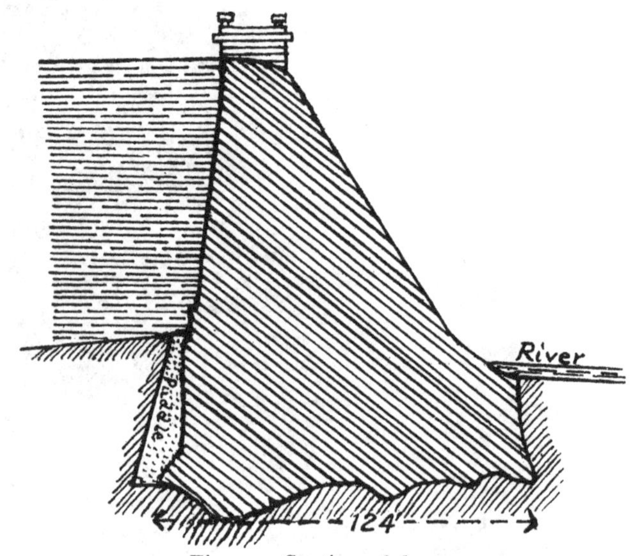

Fig. 15. Section of dam.

When a masonry dam is built across a valley to retain water in a lake or reservoir a great deal of care has to be taken in the design and construction of the dam to make certain that the pressure of the water will not overturn the dam or any part of it or push it forwards and break it across. Not only is it necessary to secure that the whole dam shall not be turned over but also that there shall be no risk of the upper part being turned over by breaking a horizontal joint. With a dam 1000 feet long and the water 60 feet deep the pressure is equal to the weight of 30 × 60,000 cubic feet of water, or 50,000 tons, which tends to push the dam forwards and it must be so heavy that this pressure will not make it slide. The resultant pressure acts

20 feet above the bottom of the water and tries to turn the dam over. The dam must therefore be heavy enough and thick enough to resist this turning action. These requirements are best met by making the dams approximately triangular in section. Fig. 15 shows the section of the Vyrnwy dam. If the solid rock is some distance below the bottom of the reservoir and the dam is built up from the rock through soft soil, as at Lake Vyrnwy, the water pressure must be reckoned as acting down to the bottom of the soft soil.

A great masonry dam gave way at Bouzey, near Epinal, in France in 1898 and blocks of masonry weighing about 1000 tons were driven forward down the valley by the rush of water from the reservoir. In this case the upper part of the dam broke away from the lower portion and turned over about its front edge.

The Birmingham Water Supply

There are very many steep river valleys in the Welsh hills at a height of 500 to 800 feet above the sea. This height is sufficient to enable the water to be carried in pipes and conduits of moderate size to any of the great towns in the Midlands or even to London, and before the formation of the Metropolitan Water Board the London County Council was considering a scheme for making a number of reservoirs in these valleys more or less after the model of Lake Vyrnwy. The Council's scheme has not been carried out and London is still dependent on local sources of supply, but Birmingham obtained powers in 1892 to erect dams so as to form six reservoirs, three on the river Elan and three on the Claerwen, tributaries of the Wye, a little above their junction. The three dams in the Elan valley were erected, and the reservoirs were formally opened by King Edward VII on the 21st day of July, 1904. The reservoirs on the Claerwen have not yet been constructed, for the original scheme provided for prospective requirements for fifty years, the collecting area being 45,562 acres with a rainfall of 63 inches a year. The whole area is capable of supplying 75,000,000 gallons a day and the six reservoirs will be able to hold 18,000,000,000 gallons. The distance from Birmingham is a little over 73 miles and the conduit has a total fall of 170 feet. Of the three dams the lowest down the valley is at Caban Côch. It is 590 feet long and the water stands 122 feet above the river bed, but 152 feet above the solid rock on which the dam is built so that the masonry has to withstand the pressure of 152 feet of water with the help only of 30 feet of earth in front of it.

Fig. 16 is a sketch-map showing the relative positions of the lakes selected by the City Council of Birmingham, on the advice of their engineer, Mr James Mansergh. The letter *A* indicates the position of the great dam and the broken line shows the first half-dozen miles of the aqueduct. Lake Vyrnwy is about 30 miles to the north. Fourteen lakes which have been recommended for the supply of London, see p. 45, lie to the south at distances varying from seven to about 20 miles, on the watersheds of the Wye and the Usk, and a little south of these and

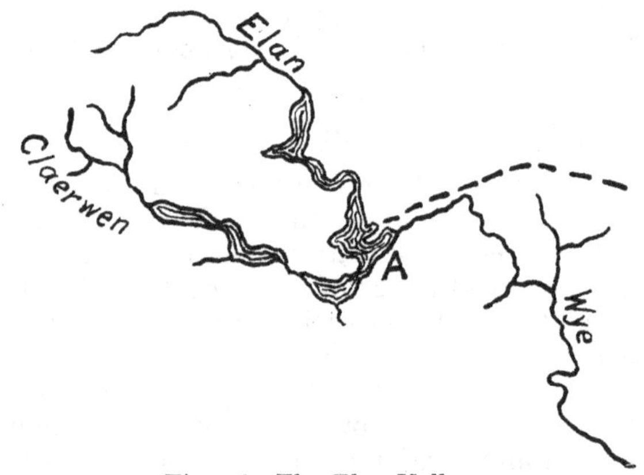

Fig. 16. The Elan Valley.

south of Brecon are the lakes which supply Cardiff, Merthyr Tydfil and Swansea.

Fig. 17 shows the distribution of the total rainfall over England and Wales during the year 1919. It is taken from the summary for the year published by the Meteorological Office. It will be noticed that in no part of the country was the rainfall less than 20 inches, though at Southend it was only 20·9 inches. There are two black dots on the sketch-map, one a little south-west of Thirlmere and the other close to Snowdon, where the rainfall was of the order of 100 inches. A glance shows that Cumberland, Westmorland and Wales are the very rainy districts and that Manchester, Liverpool, Birmingham, Cardiff and Swansea have selected gathering grounds in which they might expect a copious supply of water. It will also be noticed that

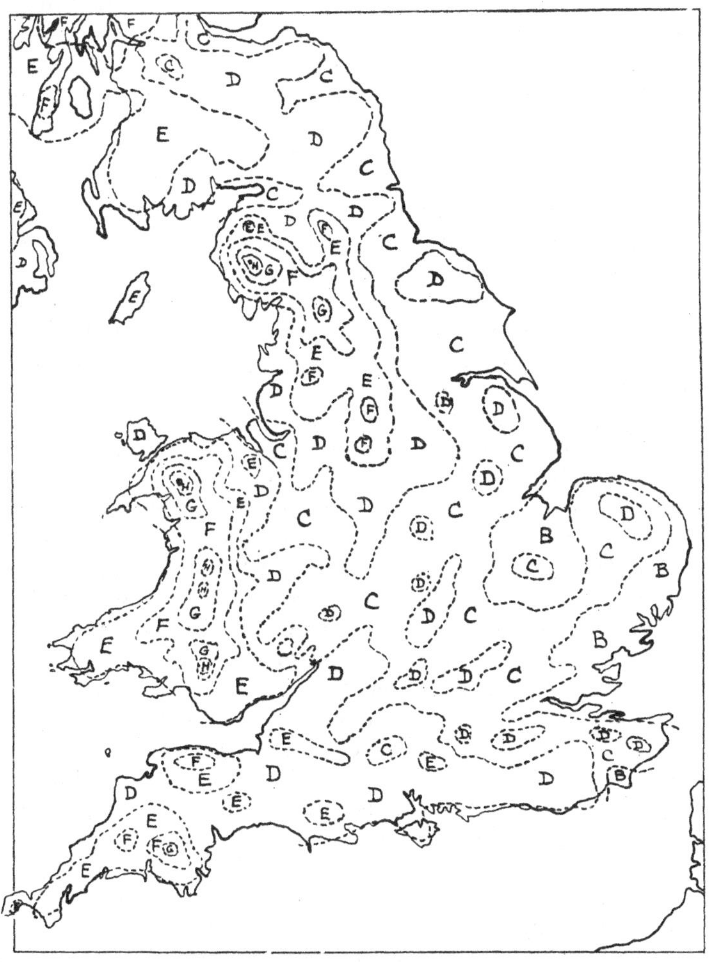

Fig. 17. Rainfall map of England and Wales, 1919.
A under 20* inches, *B* 20–25 inches, *C* 25–30 inches
D 30–40 ,, , *E* 40–50 ,, , *F* 50–60 ,,
G 60–80 ,, , *H* 80–100 ,, , ● over 100 ,,

* In no part of the country was the rainfall under 20 inches in 1919.

the greatest rainfall occurs where high mountains first intercept the south-west winds from the Atlantic, and as these mountains are formed of very old rocks which are almost impervious to water nearly all the rain finds its way quickly to the streams and lakes carrying with it very little mineral matter in solution.

Other Famous Dams

One of the biggest dams in the world is that built across the Croton River for the supply of New York (see pp. 19–21). It is 1500 feet long and 290 feet high from the lowest part of the foundation. The water-level is 163 feet above the river bed, but the masonry extends 97 feet below this. The dam which has attracted the greatest public interest of late years is that thrown across the Nile at Assouan, for by the system of irrigation which it has made possible it has rendered fertile a large tract of country previously unproductive. The dam was formally opened by the Duke of Connaught in 1902. Its length is 1¼ miles, and in this respect it vastly exceeds the Croton dam, and its height from its foundation is 130 feet. It contains over a million tons of masonry. The difference in the water-level above and below the dam is 67 feet. It raises the level of the Nile for 140 miles and impounds more than 2,250,000,000,000 gallons of water. This water can be allowed to flow out through 180 openings in the dam, capable altogether of discharging 2,200,000 gallons a second. Of these 180 openings 140 are 23 feet in height and 6 feet 6 inches wide, the others are 11 feet 6 inches high. They are intended to allow the full flood of the Nile to pass down the river with as little hindrance as possible. When this flood begins in July the river carries so much mud that if it were impounded for any length of time it would quickly silt up the river bed. All the sluices are therefore opened and the greatest flood will pass the dam with a difference of level of little more than six feet on the back and front. In November the water is clear and the sluices are closed. The water then accumulates nearly to the height of the dam and can be discharged as required for irrigation purposes in the spring. Assouan is in the district of the granite quarries from which Cleopatra's Needle and the other Egyptian obelisks and monstrous statues were quarried, and the dam is built of the local red granite. There is another dam at Assiout, 250 miles lower down the river, which controls the irrigation of Middle Egypt.

Proposals to supply London from Wales

Proposals to bring water to London from Wales had been considered and rejected by the Duke of Richmond's Commission in 1869. When the London County Council came into office in 1889 one of its first acts was to institute an enquiry into the London water supply. It soon came to the conclusion that the enquiry must be conducted by a Royal Commission, and it was on the request of the Council that Lord Balfour of Burleigh's Commission, which reported in 1893, was appointed. That Commission was of opinion that the existing sources of London water might be so developed as to furnish 420,000,000 gallons a day and so meet the requirements of London for the next forty years, allowing for an increase of the population of Water London to 12,000,000, but the Council was anxious to secure the control of the water in the interest of the ratepayers and was unwilling to pay the price which the companies were likely to ask for their undertakings. It accordingly considered the possibility of obtaining an independent supply. The Council's first scheme was to take water from the Usk and the Upper Wye and their tributaries, which were estimated to afford an average supply of 415,000,000 gallons a day. These streams are liable to be dried up entirely in times of drought so that enormous reservoirs were important features in the scheme. There were to be five of these storage reservoirs besides three compensation reservoirs. The largest was to have been on the course of the Llangorse, with a capacity of 38,000,000,000 gallons, and the next on the Yrfon with a capacity of 31,000,000,000 gallons. The reservoir on the Upper Wye was designed for 10,500,000,000 gallons and those on the Ithon and Edw for 9,000,000,000 and 4,400,000,000 gallons respectively, making a total of 92,900,000,000 gallons, or more than seven months' supply at 420,000,000 gallons a day.

The water was to be brought to London, a distance of 160 miles, in two aqueducts, one from the Usk and its tributaries and the other from the Wye. The former aqueduct was to lead to filter-beds and other works at Elstree and the latter to Banstead, whence the water would be carried in iron pipes to the present mains. Each aqueduct was to be capable of carrying over 200,000,000 gallons a day. The estimated cost was

£38,800,000. It was probable that the construction of the works would require nearly fifteen years.

In 1896 this scheme and the proposals of the water companies to construct storage reservoirs at Staines were referred to Sir Benjamin Baker and Mr G. F. Deacon, and on their advice the Council modified its proposals so as to utilise the Welsh supply as supplementary to the existing supplies, which were to be retained. Accordingly, in the Bill promoted by the London County Council in 1899 for the purchase of the water undertakings the Welsh supply was introduced as a supplementary supply, and for this purpose it was proposed to construct only one aqueduct to bring the water from the Upper Wye and its tributaries the Towey, Yrfon and Chivepri. The principal reservoir was to be on the Yrfon and over six miles in length. This scheme provided that the additional supply likely to be required during fifty years owing to the growth of London should be brought from Wales, but the present sources should be adhered to for the existing demand. This supply, like the supplies for Liverpool, Birmingham, Cardiff, etc., was to be secured by the construction of dams across rivers so as to convert the valleys above the dams into lakes or to enlarge, as at Thirlmere, lakes already existing.

Upland Sources of Water and the Thames Basin compared

When rain falls on a mountain area consisting of impenetrable rocks, like most of the Welsh hills, some of it evaporates again from the surface or is absorbed by vegetation and evaporates from the leaves, but most of the water finds its way into the mountain streams and rivers. With a rainfall of 60 inches to 80 inches a year on the Welsh hills, about 13 inches to 15 inches are evaporated in this way and the remainder flows off the surface and, if required, can be collected for use. In the Thames basin the rainfall is much less than on the Welsh hills and is only about 21 inches in a dry season and 30 inches or less in a wet season. In a dry season about 79 per cent. of the rainfall evaporates or is absorbed by vegetation, leaving only 21 per cent. to find its way to the river or to soak into the deep soil. In a wet season the evaporation is about 65 per cent. so that in the Thames Valley the rainfall is very much less than on the

Welsh hills, while the evaporation is not only relatively but actually greater. If in a mountain area there is a rainfall of 60 inches in the year and only 15 inches evaporate or are absorbed by vegetation there will be 45 inches of rainfall to find its way in the course of the year to the streams. If all this water is collected from 1000 acres it will amount to about 1,000,000,000 gallons. To supply the Administrative County of London alone and for domestic purposes only a collecting area of 70,000 acres,

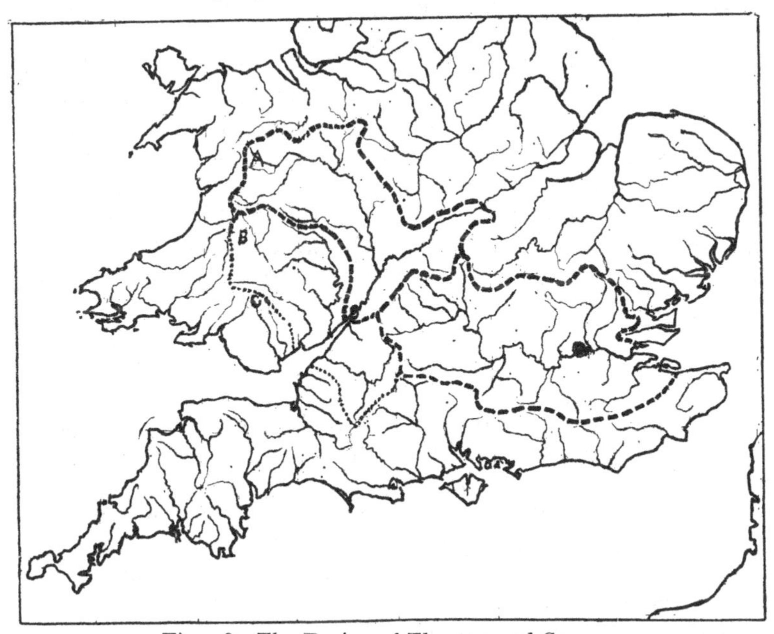

Fig. 18. The Basins of Thames and Severn.

or about 110 square miles, would be required. This is only a little less than the area of the county itself, but the rainfall in London is so much less than in Wales that if the whole could be collected it would only amount to half the demand.

Most of the rain which forms our water supply falls in the winter, for the summer rain evaporates to a very great extent and cannot be counted on to fill the rivers or reservoirs. Hence when there is a shortage of water it is usually because the winter rainfall was small. A dry summer does not make much difference to the available stores of water.

In Fig. 18 the broken lines indicate the boundaries of the basins of the Thames and the Severn, excluding the Wye and the Usk which discharge into the Severn's lower reach. They roughly indicate the water partings. To the north of the Severn basin are the basins, or watersheds, of the Trent, the Mersey and the Dee. To the south the Bristol Avon, which flows into the Severn estuary. The simple dotted lines include the basins of the Wye, the Usk and the Bristol Avon, which may be regarded as tributaries of the Severn. The letters *A*, *B* and *C* show approximately the position of Lake Vyrnwy, the Birmingham lakes and the lakes supplying Cardiff, Merthyr Tydfil and Swansea. To the north of the Thames basin are the Huntingdon Ouse and the Cam and to the south the Christchurch Avon and the smaller streams which flow into the Channel through Hants and Sussex. A comparison of Fig. 18 with Fig. 17 will bring to light the difference between the rainfall in the watershed of the Thames and that in the watershed of the Severn. If the Wye and the Usk be included with the Severn these rivers account for the greater part of the rainfall in Wales.

Springs

Sometimes a thick bed of porous rock, like chalk, or of hard rock traversed by fissures in which water can accumulate and flow will come to the surface forming high land, like the chalk hills of Surrey and Hertfordshire, so that rain water may saturate the rock to a considerable height above the sea-level. Then if there is a valley cut through the rock to a depth below the water-level the water will run out of the sloping side into the valley. If there are any cracks in the rock it will pour out, forming a spring, and the flow of water will tend to dissolve or wear away the rock and enlarge its opening. In ancient times the presence of a spring was a great attraction to primitive people to live in the neighbourhood for the sake of the water supply, and they wisely preferred springs to rivers for that purpose. After a succession of wet seasons the water in the chalk hills south of Croydon will rise so much that it will burst out into the Caterham valley forming the Croydon Bourne which will flow for some months and then disappear as the level of the water falls in the chalk hills, but a much more important outflow from the chalk is in the Lee Valley in Hertfordshire, where the Chadwell Spring frequently discharges over 4,000,000 gallons of water a day. It was this water which Sir Hugh Myddelton brought to London by the New River, but we shall learn more about this later on.

When the new-born river issues from the ice cave at the foot of a glacier something like a spring is formed on a large scale. It is true that generally the water has not travelled in underground channels but it has passed through a tunnel underneath a great mass of ice from the points at which it reached the glacier bed by flowing down the crevasses or other openings in the ice. Sometimes in the mountains water will issue from openings in nearly vertical rocks forming the boundaries of deep valleys and will fall in a cascade to the valley below. Reference has already been made to the little springs coming out of the sand seams above the London Clay on Hampstead Heath.

Possible Sources of Water Supply

It may be well here to enumerate the possible sources of water supply. They are:

1. Rain water collected on roofs and run into tanks.

2. Surface wells only a few feet deep when there is gravel or other porous soil resting on hard rock or clay.

3. Deep wells where there is a great depth of porous rock, like the greensand or chalk, immediately below the surface.

4. Artesian wells when there lies deep down a thick bed of porous rock, like chalk, to form a reservoir, with a bed of clay or other impervious rock below and another such bed above. The porous rock must have an outcrop at a higher level.

5. Springs, where the water collected by porous rocks at a high level flows through them and bursts out in the valleys or on the lower slopes of the hills.

6. Mountain rivers flowing in narrow valleys which can be dammed up to form reservoirs so as to give a steady supply of water all the year round, though the river may sometimes flow in a torrent and at other times be nearly dried up. These are included under the head of Upland sources.

7. Rivers flowing through nearly level country from which the water has to be pumped into specially constructed storage reservoirs.

Rain water collected from roofs in London and other large towns is generally too dirty for any domestic purpose.

Much of the London water supply used to be obtained from shallow wells but these became contaminated and produced epidemic diseases so that, like Aldgate pump which drew water from one of these, they have been closed.

The present water supply of London is derived from deep wells in the chalk in the valley of the Lee and in Kent where the chalk comes very near the surface without being covered by London Clay, from artesian wells which are bored through the clay into the chalk, from springs in the Lee Valley, from the River Thames and River Lee and, to a small extent, from beds of gravel on the surface generally near the river bank but remote from population.

THE DIFFERENT LONDON AREAS

Some account of the history of London's water supply will be given later on. Since 1904 it has been in the hands of the Metropolitan Water Board. There are a great many different Londons. There is the City of London with an area of a very little more than a square mile. This is the London of the Lord Mayor and Corporation. There is the Administrative County of London which for some purposes includes the City and for other purposes excludes it. This is the London of the London County Council and is also Educational London. Its area, including that of the City, is nearly 117 square miles and its population a little under 4½ millions (4,483,243 in June, 1921). Then there are the London postal district of 224¼ square miles, the London telephone area of 627 square miles, the London Parliamentary area, which differs from the Administrative County only on account of some slight shifting of boundaries, the area of the Metropolitan Water Board, known as Water London, comprising over 559 square miles with a population of nearly 7 millions, and "Greater London," which is the Metropolitan Police area together with the City and has an area of nearly 603 square miles and a population of 7½ millions (7,476,168 in June, 1921). Greater London is taken as the London District by the Registrar-General for census returns and other publications. Its boundary lies between 12 and 15 miles from Charing Cross. There are some people who think that the powers of the London County

Council should be extended so as to cover the whole of Greater London. Besides these there are other Londons for gas and electricity supply, for ecclesiastical purposes and for County Courts, Police Courts and the Central Criminal Court, so that few people, when they talk about London, know exactly what they are talking about.

The frontispiece is a sketch-map showing the boundaries of the City of London, the County of London, Greater London and Water London as supplied by the eight Metropolitan Water Companies in 1903. Tottenham and Enfield appear within the boundary though at that time they had their own independent supply. When the Metropolitan Water Board was formed the Tottenham Works and staff were taken over by the Board, but the urban districts of Ware, Cheshunt and Richmond and the county borough of Croydon arranged to provide their own supplies or to receive water in bulk from the Board, so that they are outside the Board's area of distribution. The urban district of Ware is an area of only a square mile or thereabouts a little to the north of Chadwell Spring. The other areas which went out of the district are shown on the map by horizontal hatching. Tottenham is shown by vertical hatching. The positions of the wells, pumping stations and reservoirs belonging to the Board and the general course of the New River are also shown, but the map includes some works which though sanctioned or commenced are not yet completed. Water London extends north and south from Hadlam Station to Winkhurst Green, 42½ miles, and east and west from Sunbury to Southfleet, 34¼ miles. Greater London extends to the west as far as the river at Staines.

"Water London"

It is the London water area with which this little book is largely concerned. It is made up as follows:

County of London	117	square miles
County of Essex	107	,, ,,
County of Hertford	23	,, ,,
County of Kent	159	,, ,,
County of Middlesex	97¾	,, ,,
County of Surrey	55¼	,, ,,
Total	559	square miles

Nearly two-thirds of the whole population is found within the 117 square miles of the Administrative County of London, and during office hours this proportion is considerably increased. It is clearly not possible to deal with the water supply of the Administrative County apart from the rest of "Water London," as the water supply of the whole area is under one control. This was the strongest reason why the water supply was not entrusted to the London County Council. If the Council had had authority over Greater London the story of the water supply would probably have been different. Sixty-five per cent. of the population of "Water London" is north of the river and the population per square mile within the County is eight times that of the portion of the area outside the County. Readers would do well to compare the population of "Water London" with that of Scotland, Ireland, Canada, Australia and other British dominions or the smaller countries of Europe before the war.

Sources of the London Water Supply

The sources of the public water supply of London are the River Thames, the River Lee, the chalk spring and wells at Chadwell and Amwell in Hertfordshire, brought to London by the New River, chalk wells in the valley of the Lee and in south-east London and Kent, four wells in south London and, to a very small extent, surface gravels. The Thames water, especially in dry weather, consists largely of spring water which has filtered through the chalk and oolite and rises in springs in the reaches of the river above Teddington. The numerous wells bored through the London Clay into the great chalk reservoir described below are used almost exclusively for private and manufacturing purposes and do not form any considerable part of the public supply, though they take the place of the public supply as far as the works or buildings to which they belong are concerned. These wells, as a rule, are not less than 7 inches or much more than a foot, though some are only 5 inches, in diameter. They are used not only by breweries, gas-works, public baths and manufactories but also by hotels, and large blocks of offices or other business premises, and though they do not belong to the public supply they must be taken into account in considering the peculiar situation of London with respect to the supply of water.

When it was proposed to bring water to London from the Welsh hills a Royal Commission, presided over by Lord Balfour of Burleigh, was appointed to enquire into the question, as stated above. This Commission was of opinion that there was no great city in the world so well supplied with water by Nature as London, and that to bring water from Wales would mean allowing this exceptional natural supply to run to waste. First of all there was the Thames with its collecting area of 3656 square miles above Teddington Lock, carrying not only the surface water but the spring water from the chalk and oolite and other porous rocks which occupy a great part of the river basin. Then there was the Lee bringing the water from the chalk hills of Hertfordshire and extending up to the borders of Cambridgeshire. The great spring of Chadwell and the wells at Amwell delivering chalk water into the New River, and the chalk wells in the Lee Valley and in Kent were additional sources of supply, and to these must be added the great underground reservoir in the chalk underneath the London Clay and directly underneath the greater part of London itself, pierced by something like a thousand artesian wells.

The London Basin

If you go out of London either to the north or to the south, at a distance of ten or twelve miles in the south or a little over twenty miles in the north, you will cross chalk hills which rise a few hundred feet above the sea-level. If you travel by rail you will go through some of these hills in deep cuttings and tunnels. Travelling by road you will find that you have left the clay soil of London and have come into a chalk country where there are only a few inches of dark soil in the fields above the chalk. Now this chalk, which comes to the surface in Hertfordshire and Surrey, extends underneath London where it is 600 feet or more in thickness. The top of the chalk is generally about 200 feet to 300 feet deep under London but the surface is not level. Near Regent Street it comes within 50 or 60 feet of the road and at East Greenwich it comes to the surface. Under Hampstead Heath its depth is about 400 feet on account of the little mountain of clay which is piled on top of it and forms the Hampstead Hill. The same is the case at the Crystal Palace. At the centre of the City the depth of the chalk is about 113 feet. You will see that between the chalk hills of Bedfordshire and Hertfordshire and those of Surrey the chalk forms a great basin under London and this basin is partly filled above the chalk with other kinds of earth, mostly London Clay. But the clay does not lie directly on the chalk. There is usually a bed of sand between, called Thanet sands because they form the surface soil in the Isle of Thanet. This may be of any thickness up to about 50 feet. Then come about 70 feet of clay, sand, pebbles and shells called the Woolwich and Reading beds, and on top of these comes the London Clay. Probably the London Clay at one time was several hundred feet thick all over the London basin, but most of it has been washed away. It is still about 250 feet thick at Golders Green and more than 300 feet thick under Hampstead Heath and the Crystal Palace, where the ground is high and the top of the chalk makes a downward bend. At Chiswick it is more than 200 feet thick, but in the City it is under 100 feet and it disappears altogether at the East end. The Thames flows near the middle of the basin.

Fig. 19 shows a section extending to a depth of 400 feet below sea-level from the outcrop of the chalk beyond High Barnet to the Surrey hills at Addington. The black line shows the Thanet sands. Below the black line is the chalk extending downwards for nearly 600 feet. Above it are the Woolwich and Reading beds and the London Clay. The vertical scale is about thirteen and a half times the horizontal scale, so that the slopes of the hills appear to be thirteen and a half times as steep as they actually are. Hampstead Heath is a little more than 800 feet above the datum line. In the figure *HB* represents High Barnet, *H*, Hampstead Heath, *W*, the river at West-

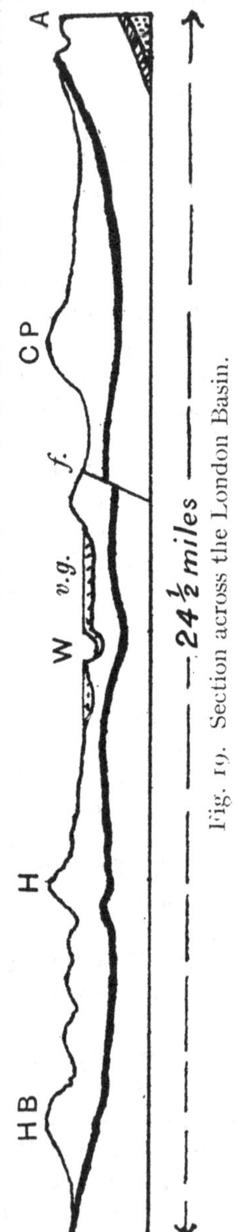

Fig. 19. Section across the London Basin.

minster, *CP*, the Crystal Palace, and *A*, Addington. Attention is called to the fault at *f* and to the valley gravel near the river. Below Addington the gault and greensand are shown.

The chalk forms London's great underground reservoir. The water flows in where the chalk and sand come to the surface in Hertfordshire, Bedfordshire and Surrey. This is 300 feet or more above the sea-level. The water flows very slowly in the chalk, but much more quickly in the Thanet sands, so that the layer of sand helps to bring the water to all parts of the top of the chalk, but the London Clay on the top of the sand prevents the water from rising so as to find its own level.

In flowing through the Thanet sands the rain water is modified in a way which would not happen by flowing through chalk only. It picks up some soda salts and oxidises the iron pyrites, or bisulphide of iron, which is found in the sands. In doing this it forms sulphuric acid, and when the water filters through the chalk the sulphuric acid acts on the carbonate of lime forming sulphate of lime while the carbonic acid helps to dissolve the chalk, so that the water from the London chalk wells drawn from below the London Clay and Thanet sands contains sulphate of lime as well as chalk and soda salts.

Where the chalk comes to the surface from under the clay and Woolwich beds it extends over an enormous area which serves for the collection of rain water in the chalk basin. In the south of London the chalk extends on the surface from Croydon to Oxted and Limpsfield, a little south of which the Weald clay is exposed. This gives a width of eight miles or more, and this width increases to the east where the chalk extends to the North Sea on the Kentish coast. Going north the chalk extends from a little south of Ware and Hertford to Bedfordshire and Cambridgeshire, a width of more than 20 miles. To the north-east it is continued to Suffolk and Norfolk. There is thus an almost unlimited area of chalk surface both north and south of London over which the rain is collected.

The sketch-map, Fig. 20, illustrates the geology of the greater part of the Thames basin and shows the chalk extending from the Channel to the North Sea on the Norfolk coast. Most of the formations are indicated sufficiently clearly in the map. The broad black line between the London Clay and the chalk represents the outcrop of the Thanet sands with the Woolwich and Reading beds, and the black patches represent isolated portions of the same beds. They extend underneath the London Clay, appearing at the surface both north and south of the clay and forming, as already stated, the easiest passage for the rain water to the lowest portion of the chalk basin under London. The gault clay, which underlies most of the chalk, is shown by diagonal hatching. It follows the lines of the North and South Downs and of the Chilterns. Between the North and South

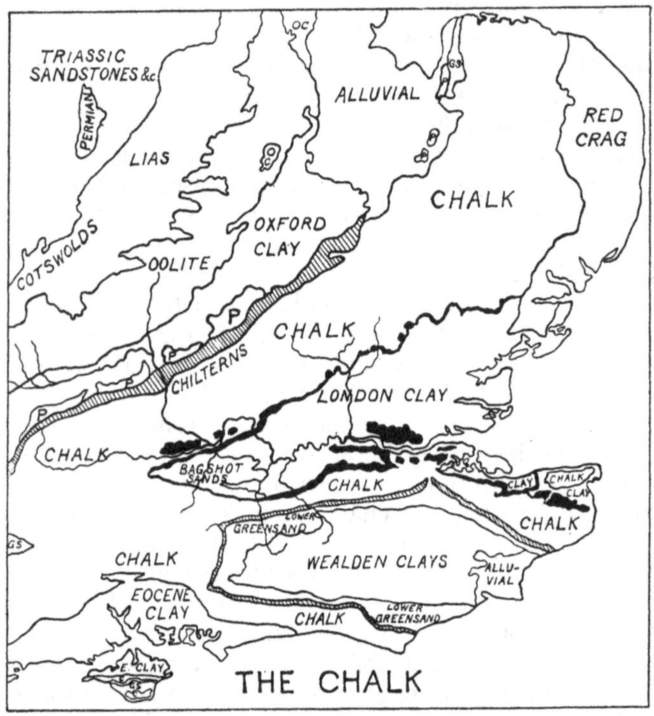

Fig. 20. Geology of Thames Basin.

Downs the chalk has been carried away exposing first the Lower Greensand and then the wealden clay in which the oak forests used to flourish. The same operation has been repeated on a smaller scale in the southern part of the Isle of Wight. The letter *P* represents Portland and Purbeck beds. In some other cases, where there is not room for the full name, initials are employed on the map, but these will be at once interpreted, *OC* for Oxford Clay, *GS* for greensand and *C* for chalk. A great part of the area of the exposed chalk is available for the collection of rain water to flow underneath the clay into London's great underground reservoir. The only rivers shown on the map are the Thames and some of its chief tributaries. The lines of the rivers will be easily distinguished from the geological boundaries.

Where the chalk is actually exposed with only a few inches of surface soil above it a large portion of the rainfall soaks at once into the chalk, perhaps 10 inches or more out of an annual fall of 28 inches. When the chalk is covered, as is frequently the case, with a few feet of drift the same happens if the drift is sand or gravel or other very permeable material, but where the drift is impermeable clay the portion of the rainfall which would otherwise soak in runs off the surface. If the surface slopes towards the chalk the water soaks in at the edge of the drift, frequently descending in "pot-holes," but if it slopes in the opposite direction the water may find its way to streams without entering the chalk.

ARTESIAN WELLS

Before any water was pumped out of the chalk for the supply of London the only course by which it could escape was by running out where the sand and chalk came to the surface at a low level in the London basin through the clay having been washed away in ancient days, or where the River Lee or some other river made a cutting down to the porous soil. Hence the chalk and sand were quite full of water whenever they were below sea-level, and on account of the resistance to the flow the water-level stood higher and higher in the chalk at points which were farther and farther away from the outflow.

Now suppose that for the first time a well is sunk near the bottom of the London Valley right through the clay into the chalk where the surface of the chalk is well below the sea-level. The pressure of the clay being removed the water in the sand and chalk will rise in the well right up to and above the level of the nearest outflow, and if the well is a long way from the outflow the water may rise considerably higher on account of the height of the hills in Hertfordshire and Surrey where the rain flows in and the resistance the water meets in reaching the outflow, so that the water may actually flow out at the top of the well. A hundred years ago this would happen in most places in the London Valley. A well bored through the clay into the chalk is called an Artesian well from Artois in France where some of the first wells of this description in Europe were sunk. Artois was an old province of France corresponding to the greater part of the present Department of Pas de Calais. An artesian well in that district has been delivering water at a height of 11 feet above the ground for over a century. Around Paris artesian wells are sunk to a depth of 2000 feet or thereabouts, and there are much deeper wells in Germany and America. Probably the deepest is at Bimerah in Queensland. The depth of this well is 5045 feet. Even in the Sahara Desert water rises to the surface in many places when a well is bored to a depth of 200 feet, and the water thus obtained makes fertile oases. In London, as we have seen, we may reach the sand and chalk at a very small depth and nowhere, except on the Hampstead and Highgate hills or the hills near Sydenham, have we to go much more than 300 feet to reach the porous water-bearing stratum.

But, unfortunately, so much water is now being drawn from London's underground reservoir that the rain cannot flow through the sand and chalk fast enough to keep up the pressure and cause the water to rise to the surface. In fact it is generally necessary to bore some distance into the chalk to get any water at all, and the water does not then rise in the well even to the top of the chalk, so that instead of flowing out at the top of the well it has to be raised by high-pressure pumps from a depth amounting sometimes to more than 200 feet. During the last sixty years the water in the London wells has in some places fallen more than 100 feet. The fountains in Trafalgar Square used to be supplied from three wells at the back of the National Gallery. They were sunk in 1847 and supplied not only the fountains in the Square but the Houses of Parliament, the Government Offices in Whitehall and Millbank Prison and could furnish 580,000 gallons a day. In 1911 two of these wells were dry, the third supplied the fountains only, but in sixty-four years the water-level in the well had fallen from 78 feet to 195 feet below the sea-level and had to be pumped up from that depth. In 1891 it was stated in evidence that the water-level in the artesian wells under London was falling at the rate of 18 inches a year. During the last year or two it has been falling at the average rate of about 2 feet a year.

If the water in a well is being pumped out rapidly the level will fall until the water flows in as fast as it is pumped out, and when the water is being pumped out of the chalk reservoir from several hundred wells at once the water-level must fall in the neighbourhood of the wells until the head of water in the chalk hills is sufficient to force a flow through the chalk at a rate to keep pace with the pumping. Hence as the number of wells increases and the water is being pumped out more and more quickly the level to which it rises in the wells becomes lower and lower, and the slope of the surface of the water in the chalk, or the hydraulic gradient, becomes more and more steep.

In some parts of the day much more water is wanted than at others so that it may very well happen that during the morning water is used much more quickly than the well can supply it, but during the latter part of the day and during the night very little water is required. When the water rose to within a few feet of the surface a reservoir was made which would hold so much water that the surface would be lowered only a few feet by the morning's pumping and would be filled up again during the night. This was done by digging the well several feet in diameter to a depth considerably below that at which the water would stand in the clay, but not down to the sand or chalk. From the bottom of this large well a hole only a few inches in diameter was bored down to the water-bearing strata and the water rose up the small bore hole into the large well from which it was pumped. The upper part of the well was carefully lined to prevent impure water from soil near the surface flowing into the well. A well of this kind is shown diagrammatically in Fig. 21. Now that the water-level in the London wells has fallen so much no attempt is made to reach it by a large dug well, and a boring only a few inches in diameter, seldom exceeding a foot, is carried from the surface down into the chalk considerably below the water-level.

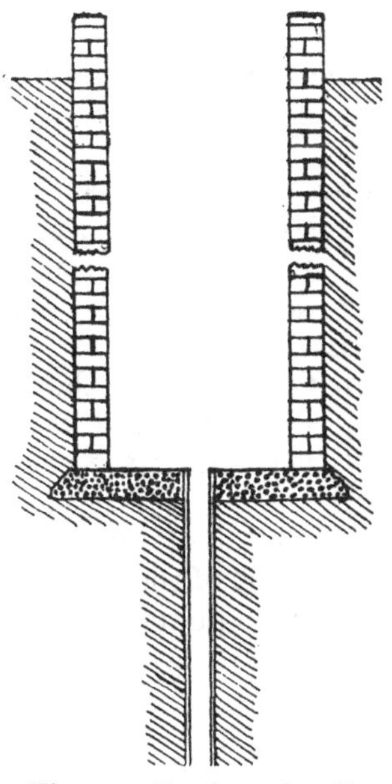

Fig. 21. Section of well.

The wells bored through the London Clay are generally from 5 inches to 13 inches in diameter, but wells of this kind are sometimes bored for public supply up to 20 inches in diameter and they may yield from 50,000 to 80,000 gallons an hour. The London borings are lined with steel tubes until the chalk is reached and then lining-tubes are generally carried some feet into the chalk. The lining-tube serves two purposes. It prevents soft soil or sand finding its way into the bore-hole and blocking it, and it also prevents impure water from the upper strata flowing into the well. When boring in some parts of the country a supply of water is sometimes found which is salt or in other ways unsatisfactory. A lining-tube is then driven into the bore-hole to a level below the sand or other seam which supplied the impure water, and the bore is continued until pure water is reached.

The amount of water which the well will yield depends on the condition of the chalk into which it is bored, for it is the difficulty which the water has in flowing through the chalk into the well that reduces the supply. When the bore-hole chances to strike a crack in the chalk below the water-level water will pour into the well at a great pace; for the crack, if it extend a long way through the chalk, will collect water from a great area of chalk wall. Thus wells very near to one another and not far from Temple Bar, although they did not vary much in diameter, gave 500, 1300 and 2000 gallons an hour respectively. These wells penetrated the chalk 307 feet, 248 feet and 265 feet and were 500, 460 and 450 feet deep. When the well consists simply of a bore-hole a few inches in diameter, if water is required during part of the day at a greater rate than the well will supply, it must be pumped into a tank, which is generally at the top of the building and is a more convenient store than the dug wells.

Where the water-bearing rock comes very near the surface, and a large well is dug until the water is reached, the collecting area is often increased by digging tunnels, or adits as they are called, from the bottom of the well horizontally into the rock like a lot of horizontal roots proceeding from a stem. This large collecting area makes it much easier for the water to reach the well, but if the adits cut across a crack in the rock a copious supply of water is at once obtained. In the case of a small bore-hole it is not possible to make horizontal tunnels at the bottom, but sometimes a cartridge of high explosive is lowered to the bottom of the well and there exploded. The shock makes cracks in the surrounding rock and may greatly increase the flow of water into the well.

Going back to the London artesian wells, in 1911 a well 10 inches in diameter was bored at Golders Green close to the Tube Railway Station. It was carried to a depth of 652 feet and passed through 259 feet of London Clay, 49 feet of clay, sand and pebbles, belonging to the Woolwich and Reading beds, 15 feet of Thanet sands and 329 feet of chalk of various kinds, sometimes mixed with flints and sometimes bearing fossils. The top of the well was 240 feet above the sea and the water rose in the well 20 feet above sea-level, so that it stood 220 feet below the surface or 39 feet above the bottom of the London Clay. Golders Green is a comparatively new district and not a great deal of water is pumped in the neighbourhood. By far the greater part of the water in this well comes directly from the Thanet sands.

A well in Bishop's Avenue, Finchley, only 7¼ inches in diameter, was bored in 1900 and passed through 375 feet of London Clay, 55 feet of clay, pebbles and sand belonging to the Woolwich and Reading beds, 17 feet of Thanet sands and 123 feet of chalk to a total depth of 570 feet. The well started at a height of 320 feet above the sea and the water rose in the well till it was 310 feet from the surface or 10 feet above sea-level and 65 feet above the bottom of the London Clay. The yield was 3000 gallons an hour.

It is interesting to compare this with a 5-inch bore-hole made in 1904 in Bunhill Row, Finsbury, to a depth of 450 feet 6 inches. This well yielded 5000 gallons an hour. The surface was 64 feet above sea-level and the water rose to a depth of 156 feet, or 92 feet below sea-level. The London Clay was reached at a depth of 27 feet 9 inches after passing through 18 feet of gravel, but in contrast to the 375 feet of London Clay at Bishop's Avenue the thickness here was only 43 feet 8 inches. Then came 51 feet 6 inches of Woolwich and Reading beds, 34 feet 6 inches of Thanet sands and the chalk was reached at a depth of 157 feet and penetrated for a distance of 293 feet.

A bore-hole 12½ inches in diameter was made at the Shoreditch Electric Light Station in 1900 to a depth of 354 feet 6 inches and yielded 3000 gallons an hour. The London Clay was only

6½ feet below the surface and 60½ feet thick. The chalk was reached at a depth of 154½ feet and penetrated for a depth of 200 feet.

When we get as far east as East Greenwich and Woolwich the London Clay disappears altogether. There is commonly river gravel just below the surface, then the layers of clay, sand and pebbles which make the Woolwich and Reading beds extending for perhaps 70 feet, then the Thanet sands and the chalk. Chalk wells in this district are not artesian wells, for all the soil down to the chalk is porous so that the water does not rise in the bore-hole. A well at East Greenwich made for the South Metropolitan Electric Light and Power Company, 12 inches in diameter and 300 feet deep, reached the chalk at a depth of 156½ feet and yielded 15,500 gallons an hour.

A few wells have been bored in London quite through the chalk and several hundred feet into the strata which lie underneath it. These borings show that below the chalk there is a thick bed of gault clay; this prevents the water in the chalk from sinking any lower. The chalk, with the Thanet sands and Woolwich and Reading beds, thus forms a porous layer under London with a thick bed of clay underneath it and another bed of clay above to prevent the water rising out of the porous layer until a hole is bored through the upper bed of clay. When the thick bed of gault clay underneath the chalk has been pierced another lot of rocks which are porous and contain water is reached. These are very much older than the chalk and the bedding is often very far from horizontal. In fact, there are very many thousand feet of rocks missing between the gault clay and these old rocks which form the ancient floor and belong to the Old Red Sandstone period. A well-boring at Chiswick has been taken to a depth of 1300 feet. The London Clay was reached at 30 feet and was 180 feet thick. Then followed the Woolwich and Reading beds for a thickness of 78 feet. The chalk was reached at 288 feet and was 655 feet in thickness. Then came 177 feet of gault clay making the waterproof bed on which the chalk rested. A thin layer of phosphatic nodules followed this and then came the marls, sandstone rock and shales of the Old Red Sandstone extending from the depth of 1120 feet to the bottom of the boring. Though the water from the chalk in a neighbouring well rose only to 181 feet below the surface the water from the Old Red Sandstone rose through the chalk and London Clay and overflowed at the top of the well when no pumping was going on, but the water contained more salt than the chalk water.

At the Beckton Gas Works the bottom of the gault was reached at 975 feet. At Chiswick, as stated above, the depth was 1120 feet. At the Great Western Generating Station at Park Royal, near Willesden, it was 1163 feet, or 1050 feet below sea-level. At Stonebridge Park, near Willesden, the depth was 1098 feet or 970 feet below sea-level. This boring has been taken to a depth of 2000 feet. At Southall the Old Red Sandstone was reached at a depth of 1135 feet or 1040 feet below sea-level, so it seems that under the whole of the London area there is a nearly level platform of Old Red Sandstone at a depth of about 1000 feet below sea-level and on this the gault, the chalk and Woolwich and Reading beds have been laid, with the London Clay above them. For particulars of the deep well at Streatham, see page 104.

To summarise, where there is gravel or sandy clay above the London Clay water can sometimes be obtained in shallow wells, but in the London area it is seldom pure enough to drink and most of the wells, like Aldgate Pump, have been closed. In the chalk and other porous beds between the London Clay and the gault clay is the great reservoir which is tapped by wells from 300 to 600 or 700 feet deep. The water from this reservoir would rise above sea-level in the well-borings if it were not kept down by pumping from so many wells. Then there is another reservoir which lies below the gault and has only just been tapped by wells more than 1000 feet deep. The water from this reservoir contains a good deal of salt, which would perhaps be washed away if there were sufficient pumping, and it rises at present to the surface of the ground.

Boys and girls who study geography learn about contour lines drawn horizontally on the earth's surface at stated intervals above sea-level, say, every 10 feet or every 100 feet, and they know that the slopes of hills and valleys can be found from contour lines shown on a map. An illustration of contour lines has already been given in Fig. 3. There have been so many wells sunk about London that it is possible to draw contour lines passing through the points at which the water surface stands in the wells which are bored into the chalk. If these lines are drawn on the map for, say, every 25 feet, they mark out the form of the surface which the top of the underground water takes in consequence of the constant pumping in the populous parts of London and the resistance to the flow of the water through the chalk. Fig. 22 shows these contour lines for every 25 feet from Stanmore to Orpington for the year 1913. The lines corresponding to odd multiples of 25 feet are incomplete in the south of London through lack of sufficient data. In the figure the principal feature is, of course, the Thames. The County

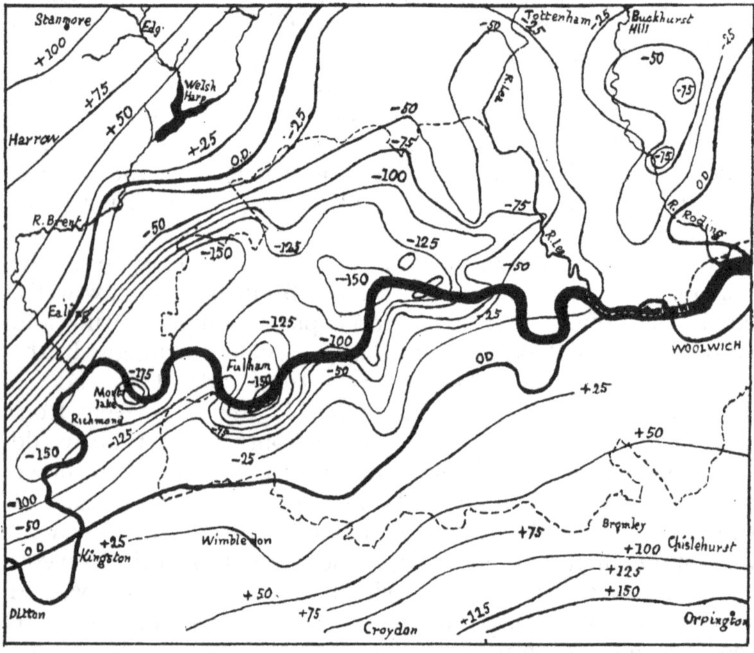

Fig. 22. Contour lines on the water gradient in the London Chalk, 1913.

boundary of London is shown by a broken line except where it is formed by the River Lee. Sea-level, ordnance datum, is indicated by the letters *OD* and represented by a line double the thickness of the other contours. The contour lines are marked in plain figures with the corresponding height above (+) or depth below (–) ordnance datum. The deep depressions in the water-level appear like islands, or mountain peaks. The deepest is at Mortlake where the inner ring, which is not numbered in the figure, corresponds to 200 feet below ordnance datum, while the inner ring near Fulham represents a depth of 175 feet. Attention is also called to the very steep gradients south of the river opposite Fulham (in the neighbourhood of Putney) and those north of the river near Brentford and Ealing. The steepness of the gradient is indicated by the closeness of the contours. Fig. 23 shows the water-level in section from Harrow to Croydon. The two steep gradients and the deep depression near Fulham are clearly shown. The lowest levels are in the middle of the valley about Mortlake, Richmond, Fulham, Charing Cross, Holborn and Cannon Street, where the water does not rise to 150 feet below sea-level. It reaches sea-level on the north side of the valley near Brentford, Ealing, Willesden, Hampstead, Highgate and Finchley. On the southern side near Hampton, Wimbledon, Tooting, Dulwich, Greenwich Park and Plumstead.

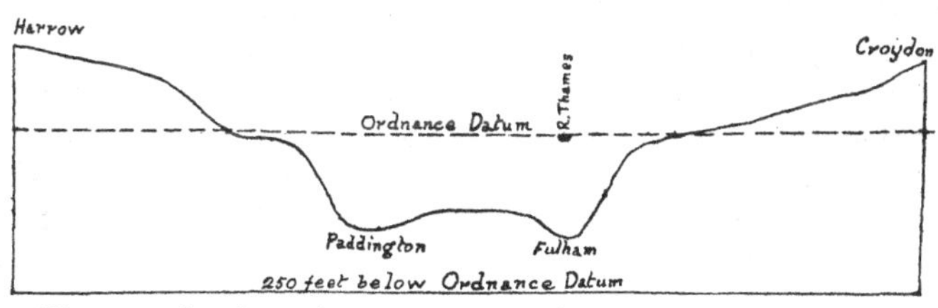

Fig. 23. Section of water contour in London Chalk, 1913.

A little south of Croydon and at Orpington water is found in the chalk at 150 feet above sea-level.

Particulars of nearly a thousand deep wells in the London area have been published. They will be found in the *Records of London Wells* by G. Barrow and L. J. Wills, published in 1913 by the Geological Survey of England and Wales. Most of the particulars of deep wells in the preceding pages have been taken from this publication. The wells are found in the middle of the City of London and in many other very unlikely places. The system of charging for water on the rateable value of large City buildings, which the Metropolitan Water Board is obliged by its Act to adopt, has led to the sinking of many wells in order to render the buildings independent of the public supply. In breweries and other works and institutions where very large quantities of water are required, as in electric generating stations or public baths, sinking a well and pumping are less costly than the payment for water from the public supply by meter, so that either a high rateable value with a relatively small consumption of water for domestic purposes or a large consumption of water for factory purposes makes it economical to sink a well. The first well sunk to the Old Red Sandstone was at Meux's Brewery, Oxford Street.

Just as the chalk lies underneath the Thames at London and rises to form hills in Surrey and Hertfordshire so the chalk which forms the Downs in the south of Hampshire dips underneath Spithead and the Solent and comes to the surface again in the range of Downs extending from the Culver Cliff to the Needles in the Isle of Wight. Tertiary clay and sand partly fill up the basin formed by the chalk. When forts were built in 1868 on the Horse Sand, the Noman Sand and the Spit Bank at Spithead wells were bored through the sea bottom inside the forts and a supply of fresh water was obtained just as in the London wells which pierce the clay and enter the chalk.

In Fig. 23 the curved line represents the free surface of the underground water on the horizontal and vertical scales adopted in the diagram, but in order to get a clear idea of the hydraulic gradient it must be observed that the vertical scale is about sixty-three times the horizontal scale and all the gradients are, therefore, about sixty-three times as steep in the diagram as in nature. The land surface is not shown in the figure. Nowhere does it fall below ordnance datum except in the river-bed. Harrow and the south of Croydon are more than 200 feet above ordnance datum so that the land surface is more than 100 feet above the water. About five miles to the south-east of Croydon the chalk hills rise above 500 feet.

Deep Well Pumps and Air Lifts

A special form of pump is commonly employed for raising the water from the bored wells. The simplest consists of a barrel with a valve at the bottom called the foot valve. This valve opens upwards and admits water to the barrel but will not let it run out again. Fitting in the barrel is a piston, commonly called the bucket. This also has a valve, called the bucket valve, which opens upwards and will not let the water run back. When the bucket is drawn up the barrel water enters through the foot valve to fill the space if the barrel is, as is usually the case in deep well pumps, below the level of the water. When the bucket is lowered the water in the barrel passes through the bucket valve to the top of the bucket, and when the bucket is raised again this water is lifted with it. If the pump is 200 feet below the top of the well the barrel, in the best pumps, is screwed to the bottom of 200 feet of steel tubing, which is only a little smaller than the tube lining the well. This tube forms the pipe or rising main up which the water is lifted when the pump bucket is raised. The top of the tube is suspended from the frame at the top of the well which carries the crank and driving gear. The pump rod passes from the crank-shaft down the rising main to the bucket. In a small well the bucket will be raised and lowered only 8 or 10 inches. In a large well the stroke may be 42 inches or more. Fig. 24 shows in section the barrel of Messrs C. Isler and Co.'s pump, and Fig. 25 shows how it is arranged in the well. From Fig. 24 it will be seen that the foot valve is provided with a strainer to keep solid bodies from fouling the valves and that it seats itself automatically in a conical seat, where it is held down by its weight. It will also be noticed that the pump barrel is a little smaller than the tube above, which forms the rising main, so that the bucket can easily be drawn to the surface for repairs. If anything goes wrong with the foot valve the bucket is raised and unscrewed from the pump rods which are then lowered again into the barrel with a guide attached to secure that the screw at the end of the rods is central. The pump rods are then screwed into the socket above the foot valve and the valve can then be raised to the surface. Fig. 25 shows the rising main a little smaller than the tube lining the bore-hole. In this illustration wooden pump rods are shown, but iron rods are more generally used. The pump barrel may be below the water surface in the well, so that the pump is "drowned," or it may be a few feet above the surface as in the common suction or "jack" pump.

In all running machinery it is desirable to maintain the load as uniform as possible to secure smooth working and avoid unnecessary shocks and wear. This is especially important when electric motors are used for driving. The pump rods with the plunger to which the connecting rod is attached and which fits into the top of the rising main, the delivery pipe leaving the main a little lower down, form a heavy load to be lifted with each up-stroke of the pump; in addition there is the column of water having the bucket for its base and, perhaps, 200 feet

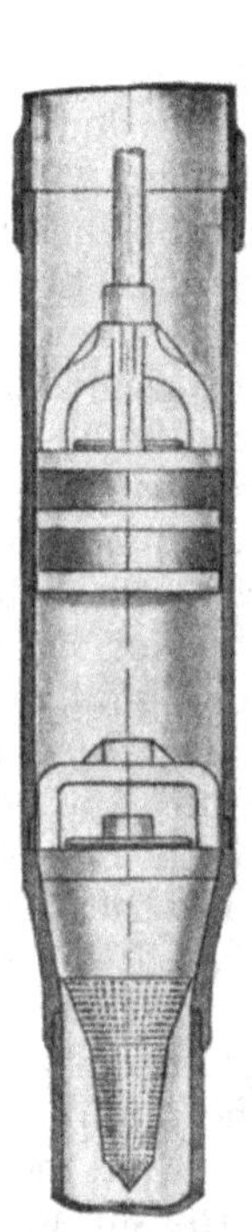

Fig. 24. Deep Well Pump Barrel, Bucket, Foot valve and Strainer (Isler & Co.)

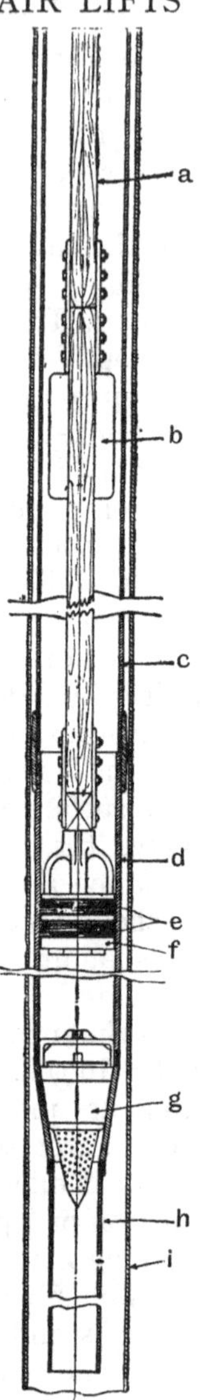

Fig. 25. Isler's Deep Well Pump. *a*, Pump rod; *b*, Rod guide; *c*, Rising main; *d*, Pump barrel; *e*, Bucket leathers; *f*, Bucket; *g*, Foot valve and strainer; *h*, Suction pipe; *i*, lining-tube.

or more in height. During the down-stroke the weight of the plunger, pump rods and bucket assist the engine. To render the driving power more uniform it is common to construct a second crank on the crank-shaft and to suspend from this crank a weight equal to that of the pump rods, etc., and half the column

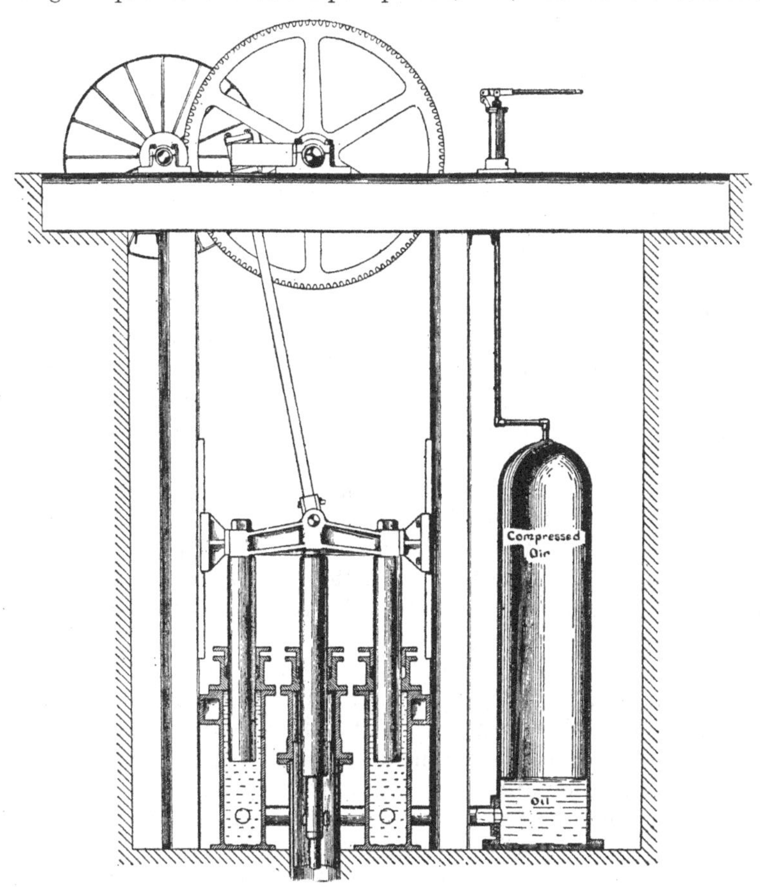

Fig. 26. Pump with compressed air chamber. (Isler & Co.)

of water. The principle is the same as that of the counterpoise on a passenger lift. With this arrangement the drive is the same during both up- and down-strokes. The reader should calculate for himself the weight of the water to be lifted by, say, an eight-inch bucket 200 feet below the point of delivery. In a few cases there are two bore-holes sufficiently near for the pumps to be driven from one motor with one crank-shaft, and by placing the driving cranks opposite to one another the pump rods may be made to balance. During half a revolution the water will be lifted in one bore-hole and during the other half in the other. But it is not often that two bore-holes are sufficiently near together to make this arrangement possible, and Messrs Isler have introduced a system of balancing which is illustrated by Fig. 26. To the top of the plunger is attached a cross-head running between guides and receiving the connecting rod, as in an ordinary steam engine. To this cross-head are attached two pistons which rise and fall with the plunger and pass through watertight collars into hydraulic cylinders which communicate with a vessel of compressed air. Oil occupies the cylinders and the lower part of the air vessel, and the compression of the air causes it to exert an upward pressure on the pistons. The pressure of the air can be regulated so as to support the weight of the gear and any desired portion of the water column. During the down-stroke the oil is forced into the air vessel from the hydraulic cylinders, increasing the air pressure, and the weight of the pump rods, plunger, cross-head, pistons, etc., helps this operation. During the up-stroke the oil transmits the air pressure to the pistons and lifts the gear and part of the water column, if the pressure has been so adjusted. The principle is the same as that adopted in connection with heavy artillery for utilising the recoil of the gun in order to bring it back into the firing position.

Air Lifts

There is another way by which water is raised from deep wells, but to make it practicable in a single lift the water must stand more than halfway up the bore-hole. Two pipes are lowered into the well. They are connected at the bottom and the large pipe is open for the water to enter. The arrangement is represented diagrammatically in Fig. 27, but to secure the best results the air on entering the open pipe must be broken up into small bubbles by a perforated screen or otherwise. This is effected by a special footpiece which can be seen connecting the two pipes in Fig. 28. Suppose that the surface of the water is 100 feet down the well. The pipes are lowered for more than 200 feet, say, 220 feet. Then the water will stand in the pipes 120 feet above the footpiece. Air at high pressure, from a rotary compressor generally driven by an electric motor, is forced into the small pipe until it has driven out all the water down to the foot-piece. It then bubbles into the larger pipe in which there is a column of water 120 feet in height. For this purpose the air must be delivered at a pressure sufficient to support 120 feet of water with a little to spare, that is to say the pressure must exceed 52 lbs. on the square inch in this case. As the air bubbles mix with the water in the larger pipe the average density of the mixture decreases so that the water rises in the pipe and at length the pressure of the water in the well forces the mixture of air and water through the delivery pipe at the top of the well. When the average density of the mixture is reduced to half that of water the pipe will deliver at 20 feet above the surface. As the column of air and water rises the pressure on the air diminishes and the bubbles expand, further reducing the density of the mixture. Fig. 28 is an imaginary picture of the bubbles expanding in the rising column and of the jet of mixed air and water issuing from the delivery pipe.

Fig. 27.

Fig. 29 represents a well similar to that shown in Fig. 21. Originally the water rose through the bore-hole into the dug well and a three-cylinder pump was employed to raise the water to the cistern. This pump with its driving gear is shown in the diagram. In course of time the water-level in the well sank below the suction of the pump so that the supply failed. An air lift, as shown in the figure, was inserted in the bore-hole by Messrs Isler, and the compressor attached to the driving gear of the pump, the air lift delivering in the dug well a few feet above the bottom. This restored the supply and the original pump raised the water to the cistern. If the depth of water in the well is insufficient for a single air lift to raise it to the surface two or three air lifts may be employed in succession on the series principle illustrated by Fig. 29, the pump being replaced by another air lift.

The Thames Supply to London

Of the total water supply of London the Thames furnishes about 59 per cent. and the Lee $23\frac{1}{2}$ per cent., leaving only $17\frac{1}{2}$ per cent. to be furnished by the springs and wells and the

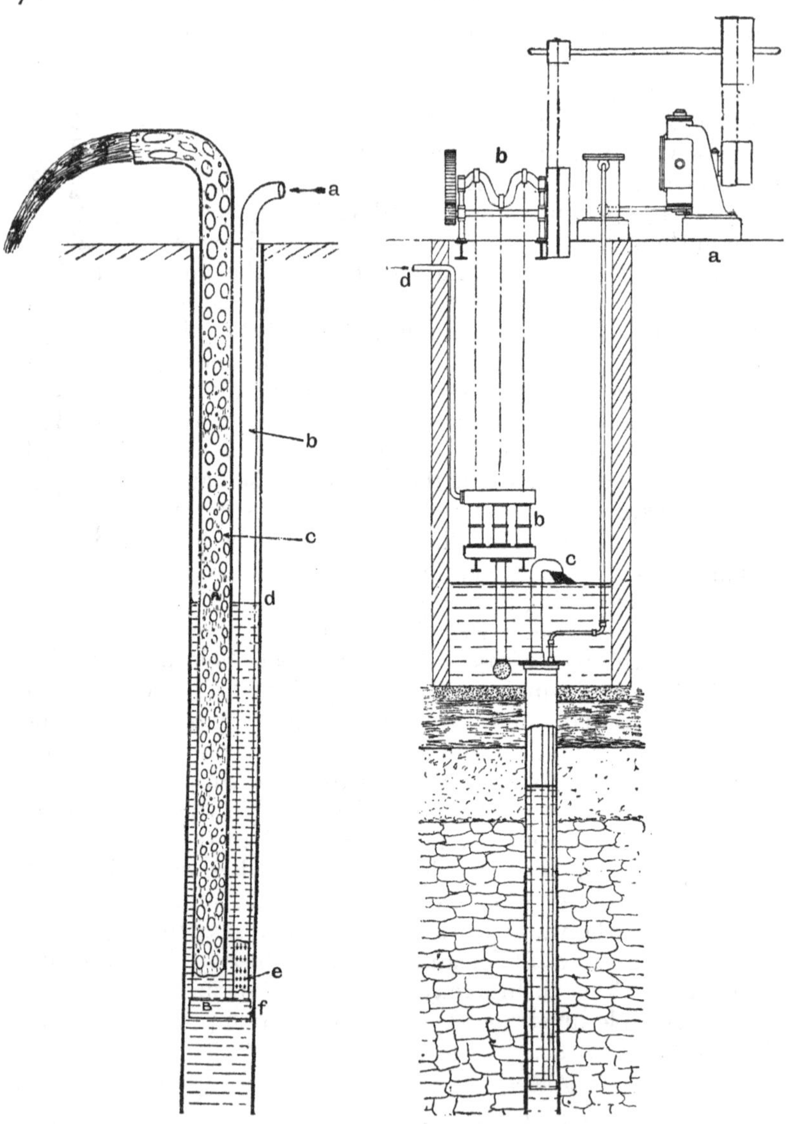

Fig. 28. Showing the action of the Air Lift. *a*, From air compressor; *b*, Compressed air pipe; *c*, Aerated water rising; *d*, Water level in well; *e*, Compressed air; *f*, Foot-piece of air lift.

Fig. 29. Isler's Improved Air Lift Pump adapted to the old system of Dug Well and Bore-hole in which water-level has dropped below suction of three throw pumps. *a*, Air compressor (belt driven); *bb*, Three throw pump (belt driven); *c*, Air lift delivery into well; *d*, Pump delivery to storage.

gravel at Hanworth. A little over 56 per cent. of the population are supplied from the Thames and 25 per cent. from the Lee, while less than 19 per cent. are provided with water from the springs and wells. As 59 per cent. of the water is distributed over little more than 56 per cent. of the population it seems that those who are supplied with Thames water get rather more than their share, in other words, the East End gets less water per person than the West End, but the difference is not great.

It follows that the chief source of supply for London is the River Thames. The water is taken entirely from the reaches above Teddington Lock. This was made compulsory by an Act of Parliament in 1852. The care of the river above the New Lock at Teddington is entrusted to the Thames Conservancy. They receive the money paid by boats passing through the locks, but the Metropolitan Water Board paid to them £40,000 a year raised to £45,000 in 1920 and to £112,500 from January 1st, 1921, in return for which the Board has the right to take water from the river up to 230,000,000 gallons a day on the average, but the Board must not take more than 1,200,000,000 gallons in any one day. The duty falls on the Conservancy to see that the river is not polluted by sewage or manufactures and this means that inspectors have to keep a watch over nearly 4000 square miles of country.

The payment made to the Thames Conservancy by the Water Board enables that body to do a great deal, not only to prevent the pollution of the river as a source of public water supply but also to make it a pleasant resort for holiday keepers, including the maintenance of the river swans as one of the amenities.

During the year 1920–21 the total amount of water which came down the Thames was 535,637,500,000 gallons, of which the Water Board and other authorities abstracted 61,028,000,000 gallons or 11·39 per cent., an average daily abstraction of 167,200,000 gallons, leaving an average daily flow over Teddington Weir of 1,300,300,000 gallons. As a rule the total flow of the Thames varies from about 600,000,000 to 6,000,000,000 gallons daily. On November 18th, 1894, the flow over Teddington Weir reached 20,175,700,000 gallons, while on July 17th, 1921, in the midst of the long drought, it broke its record and only 17,900,000 gallons passed over Teddington Weir after 138,000,000 gallons had been abstracted from the higher reaches.

The total flow of the Thames represented a rainfall of 9·6 inches over its collecting area of 3855 square miles, out of a total rainfall of 27·34 inches, or 35 per cent. In the area of the Lee a smaller percentage of the total rainfall reached the river, corresponding to a rainfall of 5·86 inches out of a total rainfall of 22·72 inches, or 26 per cent. The Board took from the Lee no less than 67·8 per cent. of the total flow.

The Thames Reservoirs

There are three classes of reservoirs for which the water authorities drawing their supplies from rivers are commonly responsible. They are compensation reservoirs, storage reservoirs and service reservoirs. The compensation reservoirs are constructed on the course of, or very near to, the river. The object is to supply water to the river during a dry season when the natural flow is very small. When there is plenty of water coming down the river the reservoirs are filled and the water is allowed to flow back into the river when required. In most cases Parliament requires that water authorities drawing water from rivers shall provide compensation reservoirs, but the Metropolitan Water Board is under no obligation to provide compensation water either for the Thames or the Lee. The largest reservoirs on the Thames are those at Staines, used as storage reservoirs. They have an area of 424 acres and hold 3,338,000,000 gallons of water.

The Staines reservoirs are constructed on level ground on a bed of gravel. The gravel is only a few feet deep and then comes clay which is impervious to water. A sufficiently thick wall of brick or stone would be exceedingly costly as it would have to be built up from foundations in the clay to prevent the water flowing under the wall through the gravel and the length of wall for the larger reservoir would be not less than 4400 yards. The walls of the reservoir are therefore made of earth with a clay core to make them watertight. First of all a trench is made in the gravel and carried down a foot or two into the clay. The trench is then filled with clay which has been mixed with a little water and worked up to the consistency of plasticine or stiff modelling clay. This clay, which is known as puddle, is rammed into the trench up to the level of the surface. The clay wall is then continued above the ground, earth being packed on both sides of it so as to support it. The clay is thus always in a trench of earth. It is carried up above the highest level at which the water is to be allowed to stand in the reservoir. The section of the wall or dam when completed is shown in Fig. 30. It will be seen that the clay forms a tank, the bottom and sides being united so as to make the whole tank watertight. The bottom is covered with a layer of gravel as it is useless to dig out the gravel below the outlet, and the clay walls are protected by gravel and earth on both sides. The slope of the earth is so arranged that there is no tendency to slip.

In large reservoirs the wind is able to make waves of considerable height which strike against the dam at the leeward end. To protect the dam from the wearing action of these waves

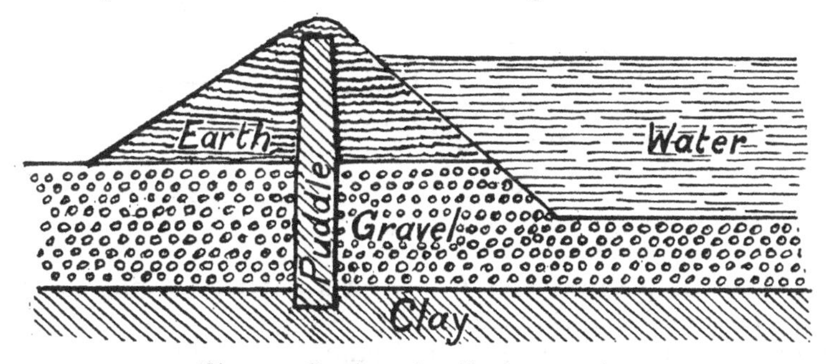

Fig. 30. Section of wall of reservoir.

it is well to cover the inside slope with large stones pressed into the earth.

There are 31 storage reservoirs for Thames water with a total area of 1026½ acres and a storage capacity of 7,267,500,000 gallons. A new storage reservoir is in course of construction at Littleton, south-east of Staines, which will have a capacity of about 6,500,000,000 gallons. Unfiltered water from the river is pumped into these reservoirs. With a demand of 164,000,000 gallons a day they can store enough water to keep up the supply for 44 days. It is in these reservoirs that the unfiltered water deposits suspended impurities and gets rid of much of its organic matter and the great majority of its living germs. When the Littleton reservoir is completed the storage capacity on the Thames and Lee together will be nearly 20,000,000,000 gallons. The present capacity of the storage reservoirs in the Lee Valley is 5,639,500,000 gallons.

The other storage reservoirs are generally constructed in the same way as the Staines reservoirs, with clay underneath them and a clay wall inside an embankment of earth. Dams for raising the level of natural lakes, by being built across the valley at the lake outfall, used to be constructed in this way before masonry dams became common in this country. The

Fig. 31.

principal storage reservoirs for Thames water belonging to the Metropolitan Water Board are at Staines (3,338,000,000 gallons), Walton (1,195,000,000 gallons), Island Barn (922,000,000 gallons), Molesey (655,500,000 gallons), Barnes (397,500,000 gallons), Hampton (390,000,000 gallons) and Kempton Park (300,000,000 gallons). For the position of these places on the river see Fig. 31 and Frontispiece.

The Thames Intakes

The places at which water is taken from the river to fill the reservoirs are called "intakes." It has already been stated that an Act of Parliament in 1852 compelled all the London Water Companies which then supplied London with Thames water to take their water from the river above Teddington Lock. The Lambeth Company had already removed its intake to Surbiton in 1848. Before this, water was taken from the river at Battersea and even as low down as Charing Cross and London Bridge, and that before the Metropolitan Board of Works had carried the drainage of London down to Barking and Crossness. It is believed that some of the epidemics of cholera were due to this polluted water supply. The Metropolitan Water Board has thirteen intakes on the Thames situated at Surbiton, Hampton, Molesey, Walton, Sunbury and Staines. Surbiton is 2·7 miles above Teddington Lock and Staines 18 miles. The other intakes are between 5·8 and 8·1 miles above the lock. These positions are shown in Fig. 31. The Sunbury intake was made by the East London Water Company, who had powers to take 10,000,000 gallons a day from the Thames, and the water was carried right across London from west to east to increase the supply in the Company's district. All the intakes, except that at Walton, were the property of the old companies before the Water Board came into existence. That at Staines was made for the storage reservoirs by the New River, Grand Junction and West Middlesex Water Companies jointly. The intakes now chiefly used are those at Staines, those at Hampton and the new intake at Walton, which was made by the Water Board itself.

The water is pumped out of the river into the storage reservoirs, or where they are excavated below the river level it is allowed to flow. Care is taken not to draw water from the river immediately after heavy rains when the river is swollen and turbid. It has already been stated that on such occasions the river carries much impurity from the roads and fields. During these times the supply of London has to depend upon the storage reservoirs, and when the river is running clear again they are refilled.

The Effects of Storage

In these storage reservoirs the water loses most of the impurities which it contains and which are not in solution, and most of the animal and vegetable matter contaminating it is destroyed. Reference has already been made to this in connection with the microbes or bacteria found in the river water. Permanganate of potash is a salt which when dissolved in water makes a solution of the colour of very dark beetroot. The colour is so strong that a single drop mixed with a tumblerful of pure water will colour the whole with a perceptible pink; but if the water contains decaying animal or vegetable substances several drops of the permanganate solution may be allowed to fall into half a pint of the water before any colour is apparent. Permanganate of potash contains a great deal of oxygen. Some of this it is very ready to give up to anything that will take it, but with the loss of this oxygen it loses its colour. Any decomposing organic matter is ready to take the oxygen which the permanganate offers it and with this oxygen it is burned up, very much as if it were burned in the air by being heated to a red heat. The quantity of a standard permanganate solution which can be put into, say, half a pint of water before any colour remains serves as a rough measure of the quantity of organic pollution it contains. If the experiment be tried with water fresh from the river and a month afterwards with water taken from the river at the same time but kept in a storage reservoir it will be found that much less permanganate is required to produce a permanent tint in the water after the month's storage. The amount of its oxygen which potassic permanganate gives up to organic matter probably depends partly on the temperature and is increased if there is any acid present.

Bacteria

The term *Bacteria* (little sticks) in common language includes, besides bacteria technically so called, many other vegetable micro-organisms which consist of single cells, multiplying generally by partition (though in some cases spores are produced which are very tenacious of life). They produce no chlorophyll (though some are coloured), and in consequence are unable to fix carbon from the atmosphere so that they are dependent on animal or vegetable food. Under favourable circumstances some of these organisms will double their numbers by partition every half hour. A single microbe may thus in thirty hours produce a colony of about 4,000,000,000,000,000,000 (4×10^{18}) individuals. As a rule they are about $\frac{1}{25000}$ of an inch in diameter; some, the cocci, are nearly spherical; others, bacteria, bacilli, spirilla, etc., are elongated. As a rule, their length does not exceed four or five times their diameter before they divide into separate individuals, but some bacilli grow to much greater length. Some of the cocci have a tendency to link themselves together in long chains, and these are known as streptococci, or chain berries. Most of these micro-organisms are harmless, and abound in air, water and in our own bodies. Many of them are highly beneficial. They produce fermentation and putrefaction and thus reduce waste organic matter to its inorganic constituents. Food which has been chemically treated, so as to destroy any micro-organisms with which it comes into contact, is very difficult of digestion. Some microbes give a special flavour to butter and cheese, and it is said that the dairies in Denmark and the cheese dairies in Cheshire and elsewhere have their special type of bacteria which they cultivate so as to secure the standard flavour of their products. Others are useful in destroying disease-producing, or pathogenic, germs. Many of the latter cannot compete with the more vigorous but harmless micro-organisms which abound in river water, and where exposed to this competition they disappear in a few days. Bacteria appear to have come down to us from ancient geological times and their remains are found as far back as the coal measures.

But while most micro-organisms are harmless and many are known to be useful there are not a few which are responsible for most, if not all, of the infectious and contagious diseases which attack men and other animals. The first bacillus to be actually convicted on indubitable evidence of the crime of producing disease was the anthrax bacillus which is the cause of splenic fever or wool-sorters' disease. This disease is common among sheep, especially in Siberia, and many other animals, and the bacillus is carried in wool and hides from which it occasionally attacked the wool-sorters and others who worked among imported wool. It has lately been found in shaving-brushes and proved guilty of manslaughter. It was in 1877 that Koch first secured the conviction of *Bacillus Anthracis*. After that many other diseases were found to be due to specific organisms which from time to time have been isolated. Cholera, diphtheria, typhoid, bubonic plague, tuberculosis, erysipelas, pneumonia, tetanus, rabies and mumps may be mentioned as a few of the diseases which have been traced to these organisms.

The disease germs most likely to be conveyed by drinking-water are the typhoid bacillus and the cholera vibrio or comma bacillus, so named by Koch because its curved form suggested a German comma. Fig. 2, p. 10, shows these organisms magnified 1500 diameters. Fig. 32 shows the pneumonia coccus and the tetanus bacillus similarly magnified. The bacillus of tetanus, which is remarkably large, is found in soil which has been treated with stable or farmyard manure and the disease is produced if the germ is introduced into the blood through a wound. Fragments of shells which had struck the ground in the highly cultivated plains of Flanders and north-east France were specially dangerous in this respect during the war. A wound from a fall on a road used by horses is sometimes dangerous from the same cause and should be carefully cleansed. Fig. 33 is an example of the chains formed by the streptococcus. This coccus

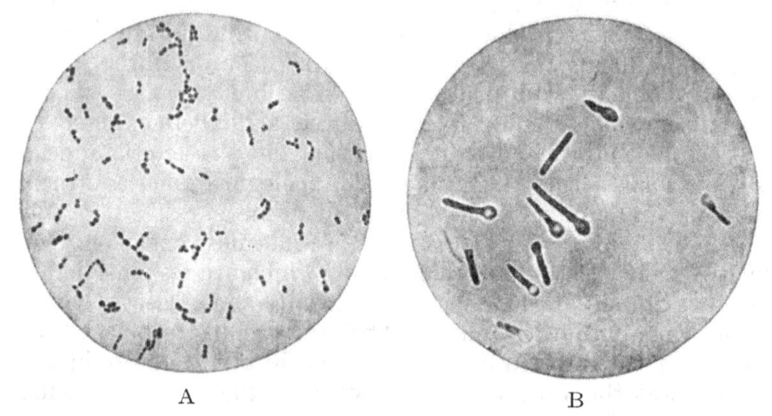

A B

Fig. 32. A. Pneumonia Cocci ×1500. B. Tetanus Bacillus ×1500.

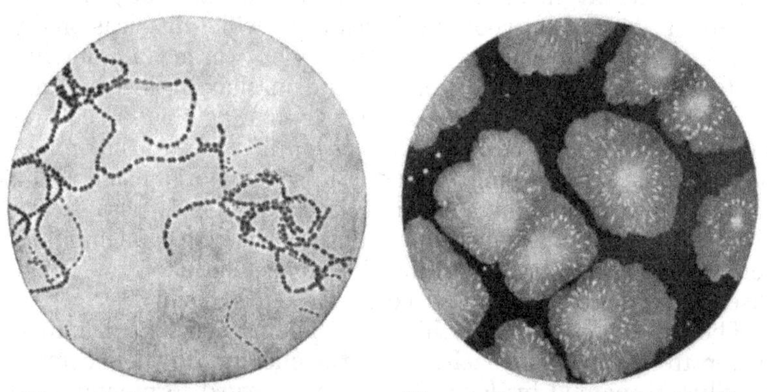

Fig. 33. Streptococcus ×1300. Fig. 34. Colonies of Colon Bacillus.

is found associated with erysipelas. Most of the micro-organisms readily absorb anilin dyes which render them conspicuous under a high-power microscope. All the illustrations of bacteria have been kindly supplied by Professor Hewlett of King's College, London. They are more fully described in his standard work, *A Manual of Bacteriology* (Churchill).

Testing for Bacteria

If a glass plate is coated with "calf's foot jelly" mixed with a little beef gravy, Lemco or other extract of meat, instead of wine, maraschino or lemon, and a little water taken directly from the river, but sufficiently diluted with distilled water, is poured on the jelly, or, better, mixed with the jelly before it has set, and this plate is kept at a temperature near blood heat for some hours, there will be found on the surface of the jelly a number of small spots, more or less opaque and sometimes exhibiting different colours. Examined by a powerful microscope it will be seen that these spots consist of colonies of bacteria. Each colony has probably sprung from one individual in the water, and the number of colonies is a rough indication of the number of active microbes in the small quantity of river water employed. Fig. 34 represents a portion of such a plate on which colonies of *Bacillus Coli* have been formed on nutrient gelatin. For some cultures agar, a gelatinous substance obtained from an Indian seaweed, is preferable to gelatin because it resists a higher temperature without liquefying.

If this test is applied to water which has stood for a month in a storage reservoir the number of colonies of microbes will be much smaller than when the water is taken directly from the river. If town sewage be employed the number will be immensely greater. If the water is taken not from a river but from a deep well in the chalk the number of colonies, if there are any, will be very small. It is now held that the most reliable record of the extent to which a river has been contaminated by sewage is the number of *Bacilli Coli* and *Streptococci* which are contained in, say, a cubic centimetre of the water. Hence considerable attention has been devoted to standard methods of estimating these numbers by cultures on nutrient gelatin and agar. A culture on nutrient gelatin is commonly prepared as described in the next paragraph.

A cubic centimetre of polluted river water may be expected to contain as many as 50,000 microbes or 50 to the cubic millimetre. In testing the number of microbes in the water we do not want to have to count more than 50 colonies, or thereabouts, but it is not possible to measure accurately a cubic millimetre of the water and work with it. It is therefore usual to add to one cubic centimetre of the polluted water 99 cubic centimetres of pure distilled water which contains no germs and to take one-tenth of a cubic centimetre of the mixture for the experiment. This will contain exactly one cubic millimetre of the polluted water. This quantity of the mixed waters is dropped from a measuring pipette into a glass dish about six inches in diameter and then some melted gelatin to which a little meat juice or similar material has been added, and sometimes a little nitrate of soda, is poured into the dish and thoroughly mixed with the water. The mixture is then spread over the surface of the dish in a thin layer and covered with a glass cover. For, say, forty-eight to seventy-two hours it is kept at a temperature of blood heat and then the colonies are counted.

The effect of Storage on Microbes

A single day's storage in a reservoir has been found to reduce the number of microbes in river water by 40 per cent. One part of quicklime added to 10,000 parts of Thames water kills all the microbes in a few hours, but this is about 33 per cent. more lime than is required to absorb the carbonic acid in the water and it leaves the water alkaline, so that it is necessary to mix it with other water containing carbonic acid but free from microbes to render it fit for domestic purposes.

The effect of storage on the life of the two microbes most likely to convey disease in river water was carefully tested by Sir Alexander Houston, the water examiner to the Metropolitan Water Board. These diseases, as stated above, are cholera and typhoid. Sir Alexander put cholera microbes into water to the number of 13,000,000 to the cubic centimetre. After one week's storage they were reduced to 20 per cubic centimetre and in three weeks they were all dead. In a similar experiment with typhoid bacilli artificially cultured, and with 8,000,000 per cubic centimetre, after one week the number was 3000, after two weeks 30, after three weeks four, and at the end of four weeks they had all disappeared. In Thames water the greatest number of microbes are to be found in the early months of the year when they may exceed 100,000 per cubic centimetre. The number is smallest in a dry summer because the Thames water is then mainly derived from the springs in the bed of the river. Of the microbes remaining in the water after storage about 98 per cent. are removed by filtering the water through sand.

Purification by Chloride of Lime

"Necessity is the Mother of Invention." During the Napoleonic wars there was a great shortage of nitre in France and gunpowder could not be made without it, so the French people had to manufacture nitre for themselves, and this they did from the nitrogenous waste products of animal life by means of nitre beds. At the same time the shortage of soda in France, when barilla, the ashes of seaweed, could no longer be imported from Spain, led to the invention of the Leblanc process of manufacturing soda crystals from common salt, an industry which for the last hundred years has flourished in Glasgow, Lancashire and on the Tyne. The shortage of sugar during the Franco-Prussian war of 1870 led to the cultivation on the continent of the sugar beet, so that for many years past our chief supplies of loaf sugar have been imported from France, Germany and Holland, a new industry having thus been created. At the time of writing (1916) it is too early to say what new industries will be developed in this country through the Great European War. We have been thrown on our own resources for many things for which we used to depend upon the continent, and in many cases we have found out how to be independent. Certain kinds of optical glass, glass chemical apparatus, hard porcelain, cheap forms of vitrified pottery, coal-tar colours and many other fine chemicals and drugs we have learned to make and, what is perhaps more important than all else, we have learned something new about the industrial ability of our population. The high price of coal due to the war inspired the officers of the Water Board to save many thousands of pounds a year by avoiding the pumping of the daily supply of water from the Thames into the storage reservoirs. This meant that some means of killing the microbes as efficient as the employment of the storage reservoir and costing much less than the cost of pumping into the reservoir should be adopted. This was found in the use of chloride of lime. Sir A. Houston's experiments showed that the addition of 15 lbs. of chloride of lime to 1,000,000 gallons of Thames water destroyed the microbes derived from previous sewage contamination far more effectually than passing the water through a reservoir containing 88 days' supply. The very small amount of chloride of lime gave no taste to the water which after this treatment was allowed to pass directly to the filter beds. About

80,000,000 gallons a day have been treated at Staines and Hampton together, with a net saving of more than £13,000 a year. It is probable that chloride of lime, like hypochlorous acid, acts on organic matter by oxidation.

Filtration

As indicated above, not all the germs are destroyed by a month's storage, and as it is most important to reduce to a minimum the risk of disease germs finding their way to the service reservoirs a further process of purification of river water, namely filtration, is adopted.

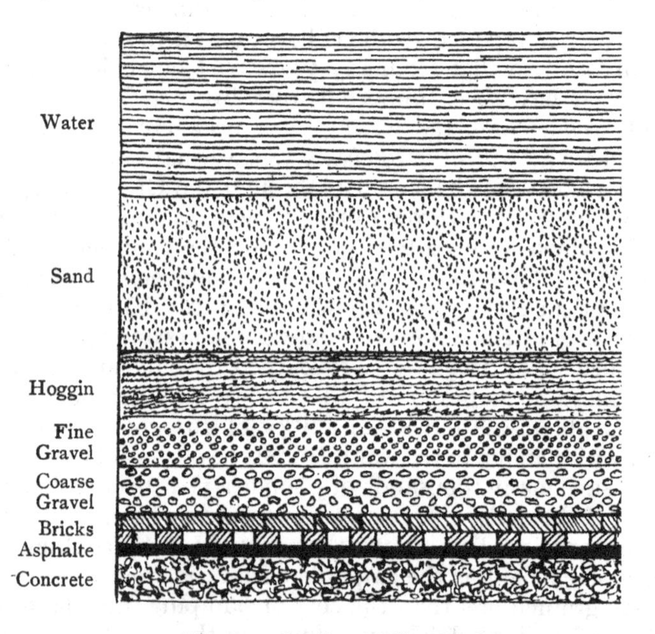

Fig. 35. Section of filter bed

The filter bed was invented in 1828 by Mr James Simpson, the engineer of the Chelsea and Lambeth Water Companies, and since its first introduction the process has been improved only in small details. Simpson's filter beds were about an acre in extent. In 1913 the Metropolitan Water Board possessed 172 filter beds having a total area of nearly 171 acres, so that the average size of a filter bed now is almost exactly an acre, like the first filter bed laid down by Simpson for the Chelsea Company. The filtration of river water was made compulsory by Act of Parliament in 1852.

To make a filter bed a tank of about an acre in area and 10 or twelve feet deep is constructed, usually with brick walls and a floor of cement concrete covered with asphalte which slopes slightly to one end, where the filtered water flows away at the floor level to the pump well. On the floor are laid rows of dry bricks, the rows being 4½ inches apart, thus forming channels of this width all leading to the end from which the water is drawn off. Across these rows other dry bricks are laid to form a continuous floor with only the narrow cracks between the bricks by which the water can flow down to the channels below. On the brick floor are laid about 9 inches of coarse gravel, then 9 inches of fine gravel followed by 12 inches of still finer, sifted gravel, known as "hoggin," and on this is placed about 2 feet 6 inches or 3 feet of fine sand. This completes the filter bed, the total thickness of the gravel and sand being 5 feet or 5 feet 6 inches. When the filter is in use the water stands about 2 feet 6 inches above the surface of the sand. A section of a filter bed constructed in this way is shown in Fig. 35.

Great care must be taken in supplying the water so gently as not to disturb the surface of the sand. This is effected by delivering the water at several points through pipes which terminate in large trumpet mouths, Fig. 36, just above the surface of the sand. The water will flow very gently over the edges of these trumpet mouths and when the filter bed is properly working the mouths will be below the surface of the water. The depth of the water above the sand is kept such that the water passes through the sand at the rate of about two gallons an hour for each square foot of surface. This

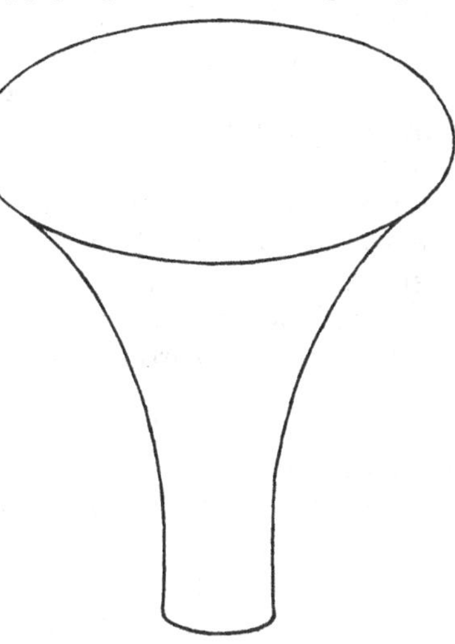

Fig. 36. Trumpet mouth.

means that if the supply were stopped the surface would at first fall about four inches in an hour as the water flowed through the sand. At this rate a filter bed of an acre would deliver about 87,000 gallons an hour or 2,088,000 a day. With 171 acres of filter beds the needs of Water London can be supplied while more than one-third of the beds are out of use.

For the first forty-eight hours or thereabouts after a filter bed has been started the filtration is very imperfect and the water which passes through it should be made to pass through another filter bed which is in good working order before it is delivered to the service reservoirs. After about two days a jelly-like (gelatinous) film is formed at the top of the sand, and it is this film which does the chief work in filtering the microbes out of the water. In the course of a month or six weeks this film becomes so thick and loaded with impurities that the water cannot pass readily through it. The water supply is then stopped and about half an inch of sand is scraped off the top of the bed. The filter is then restarted as though it were a new filter and the layer of gelatinous sand is washed in a strong stream of water which carries off the impurities and the sand can then be used again.

Service Reservoirs

From the filter bed the water, which is now fit for drinking, flows into a pump well from which it is pumped to the service reservoirs. The Act of 1852 made it compulsory that any reservoirs within five miles of St Paul's should be covered. This is usually effected by erecting a series of brick arches side by side to form the roof, which is afterwards covered with soil and grass. The arches are carried on brick piers so that the interior of the reservoir is a great hall, 18 or 20 feet high, with side walls of brick, a floor of concrete covered with asphalte and rows of pillars extending throughout to carry the arched roof. The interior of such a reservoir is shown in Fig. 37. Some of these reservoirs are erected on the highest sites available, *e.g.* Hampstead Heath, 446 feet above sea-level, Highgate Hill, 427 feet, Shoot-up Hill, 258 feet, High Beech, 371 feet, and Rock Hill, Sydenham, 365 feet. But it would involve great waste of power to raise to these heights water which is required only for low-lying districts and the pressure would be too great for the service so that other sites are selected of sufficient height for the districts to be supplied, *e.g.* Greenwich Park, 158 feet, New Cross, 163 feet, Eltham, 240 feet, Pentonville, 139 feet, and Campden Hill, 133 feet.

Fig. 37. Interior of the covered reservoir at Honor Oak.

The Metropolitan Water Board possesses altogether 86 service reservoirs. The internal area of all these reservoirs is about 69 acres, so that the average area of a service reservoir is rather more than three-quarters of an acre. The capacity of all the reservoirs is 311,000,000 gallons, so that the average depth of water, when the reservoirs are full, is about 17 feet.

Water Distribution

The distribution of water over an area like London where the levels vary by about 400 feet is a very interesting and difficult problem. As stated above, it would never do to put all the service reservoirs on the highest hills. The pressure at the lower levels would be far too great for domestic supply. Water from the Hampstead reservoir would rise above the top of St Paul's. There is in some parts of London a very high-pressure supply for working hydraulic lifts and other machinery, but this supply is independent of the Water Board and no trouble is taken to make the water fit for drinking. The pressure is far greater than any London reservoir could provide and is kept up by powerful force pumps. Another objection to placing all reservoirs on the highest hills is the great waste of power in pumping. To raise London's daily water supply of nearly 250,000,000 gallons through 400 feet would require the work of an engine of 50,000 horse-power for ten hours if there were no loss of power in the engine, pumps or pipes, and allowing for these losses we may say that an engine of 50,000 horse-power would have to work for fifteen hours a day to raise the water. At a halfpenny for a horse-power for an hour to pay for labour, coals, oil, wear and tear of boilers, engines and pumps and other incidental expenses the cost would be more than £1500 a day, and this is a pre-war estimate. At present prices it would be more than doubled. To avoid this difficulty reservoirs are placed at many different levels and the whole London area is mapped out into districts, the reservoir being placed at the highest point of each district. Then the water will flow naturally through pipes to the top of most of the houses, but there will be some houses on ground as high, or nearly as high, as the reservoir and the water will not reach the top of these houses. They can be supplied by water brought in a small main from a higher reservoir in a neighbouring district or from a standpipe or water-tower. A standpipe is a pipe of the shape of an inverted U which may be sometimes seen standing to a height of 40 feet or more above the top of a reservoir. The water for the supply of the houses at the higher level is not pumped into the reservoir but over the bend of the standpipe and this gives the extra pressure required to carry it to the tops of the houses which surround the reservoir. Standpipes similar to that shown in Fig. 38 can be seen on the reservoirs at Claremont Square, Pentonville Hill and Hampstead Heath and in many other places. The top of the bend should be open to the air, for otherwise when the air has been driven out of the pipe a syphon may be formed in which only atmospheric pressure will be reached 30 feet below the top.

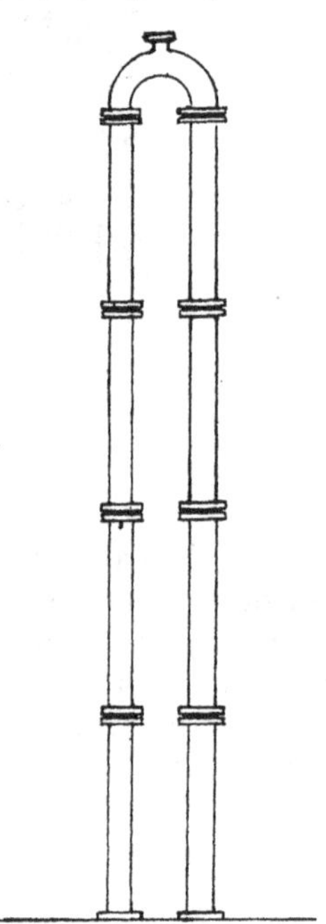

Fig. 38. Standpipe.

It is not necessary that a standpipe should be a double pipe and that the water should be pumped over the syphon. A single pipe open at the top will suffice. The sketches, Fig. 39, indicate the arrangement. The water-level, *CD*, in the standpipe rises and falls slightly with each stroke of the pump, but is always higher than the surface of the water in the reservoir or cisterns supplied by the pump. If the top of the open standpipe is expanded into a tank a considerable amount of water may be stored at the high level and will keep up the supply when the pumps are not running. The standpipe then develops into a water-tower. But merely for the purpose of maintaining the pressure and steadying the delivery it is only necessary that the capacity of the tank should exceed the amount by which the

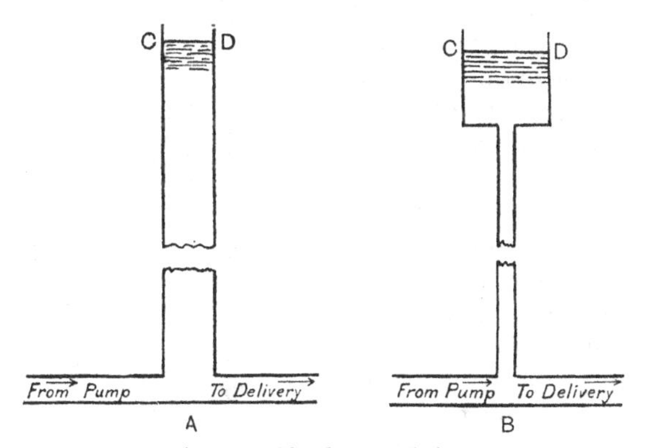

Fig. 39. Single standpipe.

water ebbs and flows in the full-sized standpipe shown in Fig. 39 A. If this is secured the pipe connecting the tank with the pump main may be comparatively small, provided that it does not offer undue resistance to the ebb and flow. With this arrangement the flow in the delivery pipe will be nearly uniform though the speed of the pump pistons is much less at the end than at the middle of the stroke. The small tank and connecting pipe are shown in Fig. 39 B.

On fire-engines and on many other pumps an air vessel is attached to the delivery tube to reduce the shocks due to the intermittent action of the pump. It is usually a pear-shaped vessel open only to the delivery pipe. The air in the vessel is compressed as water is forced in by the action of the pump and the compressed air keeps up a nearly steady pressure on the delivery pipe. It would be possible to replace standpipes by air vessels sufficiently large. A cylindrical air vessel is shown in Fig. 26, p. 69. An air vessel similar to this is employed at Hampton Works to preserve the pressure in the rising pump mains.

Suppose that in the standpipe shown in Fig. 39 A the water rises and falls 3 feet between the strokes of the pump. Then if the standpipe were cut down so as to be little more than 3 feet high and a solid plunger, fitting the pipe and weighing as much as the column of water in the standpipe between *CD* and the level of the plunger, were supported on the surface of the water, the pressure exerted on the water in the main would be the same as that exerted by the water when it stood in the standpipe with its surface at *CD*. If the plunger can be made to slide watertight up and down in the tube the water will rise and fall so as to compensate the intermittent action of the pumps exactly as in the case of the standpipe, provided there is not much friction. Moreover, the whole of the weight need not be in the plunger. By means of a suitable cross-head weights to any desired extent can be suspended from the plunger so as to increase the pressure on the water. The apparatus designed on these lines by Lord Armstrong in 1850 is the hydraulic accumulator. It is shown in section in Fig. 40, but without the guides necessary to steady the plunger in the upper part of its stroke. This plunger is made to work watertight in the cylinder by means of the well-known Bramah cup leather which is shown near the top of the cylinder and is pressed by the water against a recess in the cylinder wall on the one side and the plunger on the other. From the

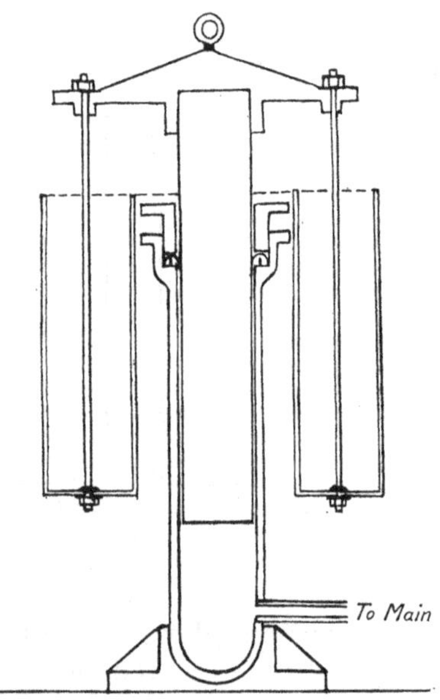

Fig. 40. Hydraulic accumulator.

cross-head, which is a circular cap, is suspended an annular tank which can be loaded with water or with any heavier material which may be available. In this way pressures up to 1500 lbs. per square inch or more may be dealt with for the transmission of power, and in these cases the accumulator may serve not only to steady the output but as a store or accumulator of energy capable of doing a considerable amount of work when the pumps have stopped, by the descent of the plunger and its load. A standpipe would have to be about two-thirds of a mile high to give a pressure of 1500 lbs. per square inch. This means that the pipes would have to be laid on a mountain side and the great length of pipe would introduce considerable resistance, but in this country mountains are not very accommodating in industrial centres. For domestic water supply no such pressures are required, but to prevent rapid changes in the motion of the great mass of water in the mains, which always means loss of energy, and to prevent shocks to pipes and pumping machinery it is very important to maintain a steady delivery, and this means some form of accumulator whether in the shape of a standpipe or air vessel or the Armstrong accumulator. Where the standpipe required is so high that a brick tower has to be erected to encase it, as at Brentford, Surbiton or Campden Hill, the Armstrong accumulator or the air compressor would probably be found more economical. In the figure the base of the cylinder is shown hemispherical because cast-iron cylinders with flat bottoms are weak along the angle between the bottom and the side on account of the crystalline arrangement of the metal.

The air vessel has the advantage over the accumulator of avoiding the friction of the cup leathers.

The total horse-power of the engines used by the Water Board for all purposes, including raising water from the river to the filter beds or storage reservoirs or from the deep wells to the pump wells, in addition to that employed in raising the water to the service reservoirs, is given as 43,910. This is less than half the horse-power available in one of the fastest Battle Cruisers. The total number of engines is 281 and of boilers 496. The engines include some of the most modern type, but many which are of historic interest. The old beam engine is still held in respect in many pumping stations, as it is very reliable, runs slowly and seldom requires repair. The most modern pumping engines are the Humphrey pumps at Chingford. These combine a gas engine and pump in one, for the engine cylinder is part of the water main in which Dowson gas and air are exploded above the surface of the water and the natural oscillations of the water serve for suction, compression and exhaust. The total coal consumption is a little more than 500 tons a day on the average. It has been pointed out that it would not pay to raise all London's water supply to the level of the Hampstead Heath reservoir, 446 feet above the sea, or to the top of Highgate Hill, but some of the districts in Kent and Surrey supplied by the Water Board require the water to be raised to a much higher level. Perhaps the highest reservoir belonging to the Board is at Betsom's Hill, near Sevenoaks. The top of the water in this reservoir is 818 feet above sea-level. The reservoir at Knockholt Beeches, near Bromley, is 795 feet above the sea, that at West Wickham 550 feet and that at Farnborough 439 feet. The tank on the land adjoining the Crystal Palace is 438 feet above the sea, but the great majority of the London reservoirs are between 100 and 300 feet above sea-level. In all cases the measurements are taken from the "Ordnance Datum" to the surface of the water when the reservoir is full.

There are a few cases in which standpipes are carried to a very great height. Such pipes are encased in brickwork and resemble tall chimney shafts. Perhaps the best known standpipe in the London district is that at Kew Bridge, though probably comparatively few people who have seen the tower only occasionally know that it contains a standpipe. In this standpipe the water is raised to a height of 184·5 feet above sea-level. The water surface in the reservoir on Campden Hill is 133 feet above sea-level, but the standpipe contained in the brick chimney tower, which is such a conspicuous feature of the Campden Hill district, enables the water to be raised through an additional 137 feet to a height of 270 feet above ordnance datum. The water is pumped to a reservoir at Shoot-up Hill where the top water-level is 258 feet above ordnance datum. A careful observer will notice window slits in the square brick tower. The standpipe is situated between the central chimney shaft and the brick casing.

There is one example in the London district of a tank constructed on a "water-tower" so raised as to give a continuous supply at a height above the highest land in the neighbourhood.

A standpipe will afford a supply only while the engines are running, but water-towers form small service reservoirs on brick (or stone) pedestals. They are very common in country districts. The only example in the London area may be found at Shooter's Hill where a water-tower was erected for the constant supply of about 350 houses which previously received only an intermittent supply. The water-level in the tank is 489 feet above the sea.

The *work* of distributing the water is done in the pumping stations where the water is forced to the service reservoirs. The engines and pumps which raise the water from the rivers or wells serve simply to collect the water. After collection it is stored and filtered if necessary. Then it is raised by force pumps in the pumping stations to the service reservoirs. After that its journey is downhill to the houses supplied and gravity does all the work. The same is the case when the water is pumped over a standpipe. The engine raises the water to the top of the standpipe and when it has flowed over the bend gravity carries it to the domestic taps at a lower level. Besides affording the necessary "head" the standpipe keeps the pressure nearly constant between the strokes of the pump.

Pumping Stations

Deptford pumping station may be taken as an example of a station taking water from the chalk wells, like all the pumping stations in the district of the old Kent company. The water is pumped from three wells which together yield about 8,000,000 gallons a day. There are three pumping engines used for raising the water. These engines could raise 50 per cent. more than the wells yield, but the raising of the water from the wells is the smaller, though essential, part of the work of the station for the water has yet to be forced to the reservoirs. For this purpose six engines are used to drive the force pumps and together they could deal with 13,000,000 gallons a day. They not only supply Deptford, which is at a low level, but can pump over 8,000,000 gallons a day to Greenwich Park reservoir, 158 feet above sea-level, nearly 3,000,000 gallons to New Cross, 163 feet above sea-level and 1,750,000 gallons to Woolwich Common, 248 feet above sea-level. There are covered reservoirs at the pumping station in which water is stored so that on some days the amount pumped to the service reservoirs may exceed the yield of the wells. The water coming directly from the deep chalk is not filtered nor is it stored for the purpose of purification.

A good example of a pumping station for river water is that at Ditton. The bulk of the water is pumped a distance of 10 miles to two reservoirs at Brixton, 115 feet above sea-level. The water is pumped through four mains each 30 inches in diameter. The remainder is pumped through two 12-inch mains to the reservoir at Coombe Warren, 180 feet above sea-level, or over the standpipe at Coombe, 220 feet above sea-level. Nearly 49,000,000 gallons a day can be pumped by engines of the aggregate of 2140 horse-power. All this water is filtered before it is pumped. The resistance of the pipe line to Brixton is equivalent to adding 65 feet to the height to which the engines are pumping.

The engines at the Hornsey Pumping Station possess 960 horse-power and are able to raise 27,700,000 gallons a day to the reservoirs supplied by them. They deal exclusively with New River water. The heights above sea-level of the top of the water in the reservoirs which are served from Hornsey are Crouch Hill 195·5 feet, Bourne Hill the same, Hornsey Lane 348·2 feet and Southgate 308 feet, but with a standpipe rising to 333 feet.

An example of pumping in series is afforded by the water sent from the Staines reservoirs to some of the higher service reservoirs in the north of London. First of all the water has to be pumped from the river into the storage reservoirs, 42 feet above sea-level. The aqueduct which brings the water from these reservoirs is tapped at Kempton Park and the water pumped through 13½ miles of 42-inch main to Cricklewood, 138 feet above sea-level. From Cricklewood it is pumped through 3½ miles of 42-inch main to Fortis Green, Finchley, at a height of 298 feet. From Fortis Green some flows by gravitation to Maiden Lane and Crouch Hill reservoirs, but some is again pumped to Hornsey Lane, at a height of 348·2 feet and some to Southgate at a height of 333 feet to the standpipe.

Iron Mains

The water is carried from the service reservoirs through the principal roads in iron mains which are pipes of cast-iron provided with an enlarged socket at one end. On the other end, which is called the spigot, is a small projection or bead which makes a loose fit in the socket of the next pipe. In laying (or "driving") the main the spigot end of the pipe is inserted in the socket of the pipe last laid and a little tow driven in against the bead. A strip of clay is then laid against the front of the socket so as to seal the space between the two pipes except for an opening at the top. Through this opening is poured melted lead until the space between the tow and the clay is quite full. When the lead is hard the clay is removed and the lead is then driven in as firmly as possible by caulking tools, or blunt chisels, struck by hammers. This makes a sound joint which will resist great pressure of water. The largest cast-iron mains in London are 4 feet 6 inches in internal diameter. The total length of mains belonging to the Water Board exceeds 6500 miles.

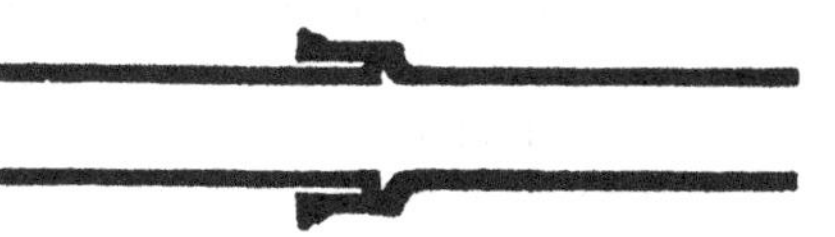

Fig. 41. Joint in iron main.

Service Pipes

From the mains are taken off iron service pipes and lead pipes branch from these to the houses. Near the point where the service pipe leaves the main there is a stop-tap or valve in the pipe which is opened or closed by a turncock. At the other end of the service pipe is a wash-out valve which can be opened so as to allow of a great rush of water into the street to scour the pipe.

The house service pipes, or connections, are of lead and these are provided with cocks in little boxes outside the houses. By opening these boxes the inspectors of the Water Board can reach the pipes and by placing one end of a rod on the pipe and listening at the other end they can tell whether any water is flowing into the house. By trying this experiment in the middle of the night they can find out whether water is running to waste in the house. In 1920–21 there were 1,134,719 services.

Waste Water and Venturi Meters

The loss of water through waste is a very serious matter in any system of water distribution and sometimes it amounts to half the total supply. It was stated before Lord Balfour's Commission that the excessive amount of water per head supplied by the Grand Junction Company was due to the company finding it cheaper to pump water than to prevent waste. Some of the waste takes place in the mains and service pipes and is outside the control of the consumers. The Venturi meter is a pressure gauge attached to the main at a point where the diameter is greatly reduced so as to increase the velocity of the stream in inverse proportion to the reduced section. Higher speed means lower pressure, according to a well-known law, and the reading of the gauge enables the velocity of the water to be determined. Service pipes are now so well laid that the principal sources of waste are found within the houses. Taps are carelessly left open or washers are worn out and the taps leak. These sources of waste are so serious that a staff of inspectors is maintained by the Water Board to find them out. The waste water meter is an ingenious contrivance which is attached to a service pipe. It was first used in Liverpool in 1873. It records on a moving strip of paper the rate at which water is flowing through the pipe. The record is something like that of a recording barometer. A pencil moves up and down according to the flow of water through the meter while the paper is moved horizontally by clockwork and the time is marked in hours and minutes on the paper. The observations are made at night when little or no water is being consumed. If an ordinary tap is opened and closed the pencil jumps up suddenly, remains steady while the tap is open and quickly falls as the tap is closed. If a ball-tap is opened the pencil jumps up and falls very slowly as the ball rises in the cistern and cuts off the water. But water running to waste goes on steadily and is shown by the pencil never falling to the lowest point which corresponds to no water passing. By examining the line drawn by the pencil it is easy to separate waste from water which is purposely drawn. When the meter shows that there is too much waste on any service pipe an inspector goes round the district in the middle of the night and turns the supply off one house after another noticing the exact time at which it is turned off each house. When he has turned the water off a group of houses he turns it on again noticing the time in each case. He then goes to the meter and examines the record. If it shows that the waste was reduced at a particular time he has only to refer to his notebook to learn the particular house from which he turned off the water at that time and he will have good reason to believe that if he calls there in the morning he will find a leak somewhere.

Wooden Water Mains

Prior to 1810 wooden water mains were in use in London and other towns, although iron water pipes made their appearance in 1746. These were generally made of elm because that wood can be exposed for a long time to water or damp earth without rotting. Elm trunks were bored out to a diameter of about 7 inches to form the waterway. Then the smaller, or spigot, end was sharpened to form a cone and at the other end the bore-hole was enlarged into a conical socket to receive the spigot end

Fig. 42. Wooden water mains crossing the Fleet at Holborn.

of the next length of pipe. The process of laying consisted in driving the spigot into the conical socket, and until to-day water engineers speak of "driving a main" while gas engineers speak of "laying" it. Wooden pipes made of pine poles, like scaffold poles, may still be seen in the Alps fixed by the side of the foot-path in a mountain valley and bringing water to a village or Kurhaus. In the case of some medicinal springs the water from which attacks iron the wooden pipes are of special value. It is said that wooden pipes with a waterway 8 feet in diameter are made from the giant trees of California. Between 1810 and 1820 the wooden pipes were generally replaced by cast-iron, but steel pipes are sometimes used where great pressure has to be sustained. Although the Chelsea Company introduced iron pipes between 1755 and 1760 they did not use them generally and the New River Company kept to wooden mains until after 1810. The pipes which bring the water to the turbines for driving the dynamos at the Simplon Tunnel and for the Jungfrau Railway are built of riveted steel plates and each length looks like the shell of a steam boiler. The Metropolitan Water Board now uses steel mains very largely at its pumping stations and where pipes of great diameter are required in connection with storage reservoirs, as at Staines.

Fig. 42 shows the condition of the wooden water mains which carried water across the Fleet River very close to the spot at which Holborn Viaduct now spans the valley.

Iron Mains and Constant Supply

In 1809 application was made to Parliament by the West Middlesex Water Company for powers to raise additional capital to carry out works estimated to cost over £157,000 and to supply water over a considerable area, including Paddington, St Marylebone, St Pancras and Bloomsbury, already supplied by the Chelsea Company or the New River Company or both. At that time the reservoir on Campden Hill, which had just been completed, was supplied by water pumped from the river at Hammersmith. The standpipe and its tower were not then in existence, but the reservoir was sufficiently high to carry the water to at least the first floor of the houses in the highest parts of the district. The proposal was to use exclusively cast-iron pipes and to keep the mains always full of water. From the evidence which was brought before the Parliamentary Committees, to which the Bill was referred, it appears that over the west and north-west of London in 1810 water was supplied to the houses for a short time only on three days in the week and sometimes it failed to come on for two or three weeks in succession in some districts. In the higher parts of the area water could only be supplied in the basements of the houses and the period of supply was so short as to encourage the practice of bribing the turncocks to keep the water on for a longer time in order to get sufficient into the cisterns to meet domestic needs. At that time 200 gallons a week, instead of 200 gallons a day, was the quantity considered sufficient for an average house. The practice then, and for many years afterwards, was to turn on the water from the mains to the service pipes, from which the house connections branched off, for a certain time, perhaps half an hour, then to turn it off from the service pipes in that district and turn it on in another district so that there were never more than a few streets in which the houses were receiving water into their cisterns at the same time. This prevented an excessive demand on the pumps. The West Middlesex Company secured the powers they sought and it seems that the Parliamentary Committees were greatly influenced by the undertaking of the company to keep their mains always full of water, night and day. This they could do through the exclusive use of iron mains. The wooden mains used by the other companies were so leaky that the water was turned off from the mains themselves as soon as the cisterns had been supplied throughout the area. If a fire occurred at night, when the fire-engines arrived the first thing to do was to remove the fire-plugs so as to get direct access to the main, but when this was done the main was usually empty. A message had then to be sent to headquarters to turn the water into the main and the engines had to wait sometimes an hour, sometimes an hour and a half, until the water appeared at the plug holes. The Chelsea Company had a reservoir in the Green Park, but this was not high enough to fill the mains, so they had a small raised tank by the side of the reservoir and a horse mill for pumping water from the reservoir to the tank. The horses were always in readiness to work the pumps, and when news of a fire reached the man in charge through the watchman in the Park, for there were no electric fire-calls in those days, he first opened a valve which allowed the water from the tank to flow into the main and then brought out his horses. Before the tank was empty the pumps were at work to keep up the supply. It was estimated that the tank itself held enough water to fill a quarter of a mile of main, but if the fire were two miles away and more than two miles of main had to be filled before a proper supply reached the fire-engines it will be realised that a long time elapsed. The damage done by the fire at St James' Palace a short time before this enquiry was largely due to the impossibility of obtaining sufficient water for the engines from the plugs. About a dozen engines were on the spot and at length by order of the Duke of York and Duke of Cambridge water was obtained from the "Canal" in the Park by arranging the engines in a series of nine or ten and passing the water on through the hose-pipes from one to another until it reached the Palace, but by this process the nine or ten engines could do only the work of one. Other cases were mentioned in evidence where a dozen houses, and in one case eighteen houses, were destroyed through the long wait for water when only one would have been burned if water had been on the spot. It is not surprising that under these circumstances the proposal to keep the mains always full from a reservoir on a hill top with a capacity of 2,300,000 gallons offered great attractions to Parliament, although a plumber of the Company told the Committee that it was easier to pass water uphill through a three-quarter inch pipe than through a pipe an inch and a quarter in diameter because the water in the pipe would not weigh so much.

In the course of the enquiry the consulting engineer of the West Middlesex Company, who was also resident engineer of the East London Company and consulting engineer to several other companies, was asked what he meant by an engine of 72 horse-power, and he replied "An engine which can lift 72 times 180 lbs. over a pulley." He explained that a horse could just lift 180 lbs. by a rope over a pulley and an engine of 72 horse-power could just lift 72 times as much. This agrees with the accepted standard of horse-power only if the engine is lifting the weight at the rate of 183½ feet a minute, and this is probably not very different from the piston speeds then common in pumping engines, but it does not justify the definition.

In 1810 the Chelsea Company were using iron mains for new work while the New River Company were keeping to the wooden mains or "trees." In 1817 an Act was passed requiring that after ten years all new mains should be laid in iron.

THE LEE VALLEY SUPPLY

Though the Lee[1] is a tributary of the Thames the basin from which it derives its water is quite separate from that which supplies the Thames above Teddington Lock, and there are many other features of the Lee Valley supply which distinguish it from the supply derived from the Thames.

The spring at Chadwell, the water from which was brought to London in 1613 by Sir Hugh Myddelton through the New River, and the wells at Amwell and Ware, belong to the Lee Valley, and besides these and the supply from the river itself

[1] The name of the river which forms the North-East boundary of the County of London is now generally spelt Lea, but in old maps and documents it appears as Lee and the Water Board preserves the old spelling.

the Water Board pumps water from no less than twenty-eight wells in the same valley.

The River Lee rises a little to the north of Dunstable in Bedfordshire, and its length is over 40 miles before reaching the point at which it discharges into the Thames at Blackwall. The Stort, which is one of the principal tributaries of the Lee, rises on the borders of Essex a very little south of the source of the Cam. The basin of the Lee is separated from that of the Bedfordshire Ouse by the Chiltern Hills, which are part of the chalk. Those readers who know the East End of London and are familiar with the Lee where it passes Stratford or is crossed by the Barking Road must not conclude that the river as they know it bears any resemblance to the upper reaches from which the water supply is taken. The Thames at Tower Bridge is a very different river from the Thames at Sunbury or Staines.

The quantity of water which the Water Board may take from the Thames in a year or in any one day is strictly limited by Act of Parliament, but the water in the Lee, like that in the New River, is the absolute property of the Water Board, subject only to the condition that it leaves enough water in the river for the purposes of navigation. This right comes to the Water Board from the former East London Water Company and the New River Company to which it was given by the River Lee Water Act in 1855.

While the Water Board pays £112,500 a year to the Thames Conservancy it pays £24,000 a year to the Lee Conservancy to meet the cost of preventing the pollution of the water by keeping watch over all tributary streams.

When the demand upon the New River Company exceeded the supply from the springs a cut was made at New Gauge connecting the upper part of the Lee with the New River so that the New River has recently brought water from the Lee as well as from the chalk springs and wells.

If the course be traced of the River Lee and its tributaries, the Rib and the Stort, on a geological map it will be seen that they all flow through a chalk district, and after their union the River Lee passes from the chalk into the London Clay a little south of Ware. It is here that the chalk dips below the clay and here that the spring of Chadwell issues from the chalk.

Fig. 43 is a sketch-map of the River Lee with its principal tributaries, and the New River as it was in 1800. The broken line represents the boundary between the chalk and the Woolwich and Reading beds which border the London Clay. It will be

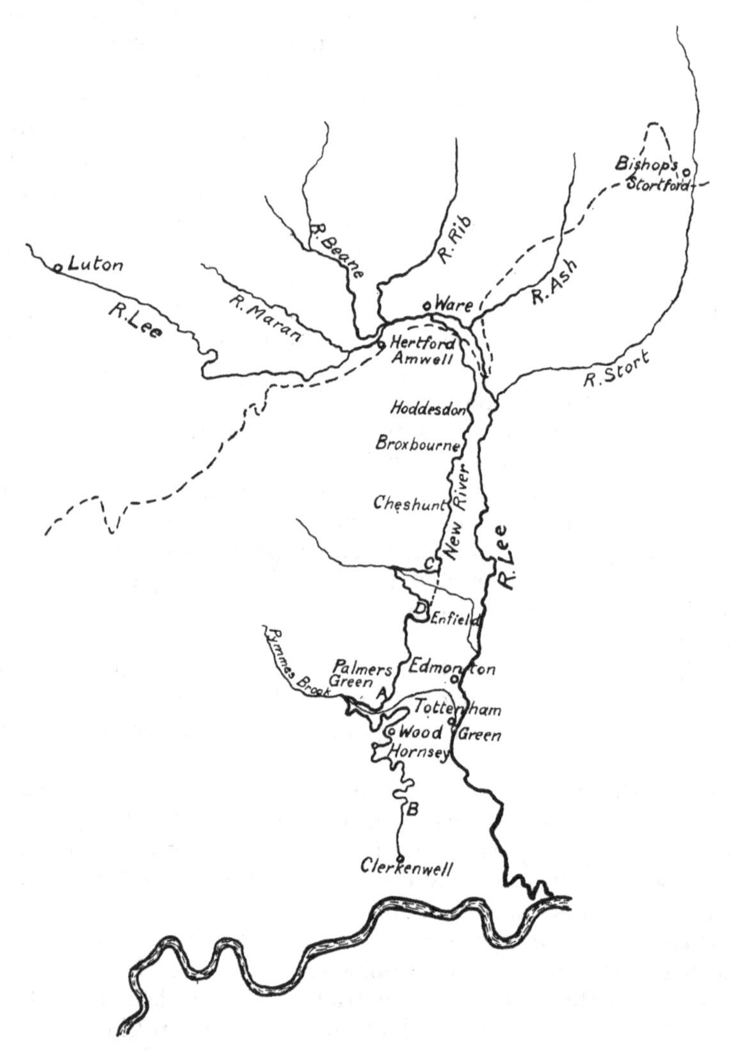

Fig. 43. River Lee and New River in 1800.

seen that all the principal tributaries of the Lee, as well as the Lee itself, flow for a considerable distance through chalk before they unite. The map does not extend northwards to the sources of most of the rivers.

The East London Company, which shared with the New River Company the right to take water from the Lee, constructed in the early fifties storage reservoirs at Walthamstow and filter beds at Lee Bridge. There are now 479 acres of storage reservoirs at Walthamstow and Tottenham receiving water from the Lee and the wells in the Lee Valley, but the largest reservoir near the Lee is that opened at Chingford by King George V and Queen Mary in March, 1913. Its area is 425½ acres and its capacity is 3,073,000,000 gallons. It cost £548,000. The small reservoirs at Stoke Newington (Green Lanes), Hornsey and New River Head (Clerkenwell) have an area together of 51¼ acres, making a total area for the Lee Valley of 955¾ acres of storage reservoirs with a capacity of 5,639,500,000 gallons. There are now thirty-one storage reservoirs on the Thames and seventeen on the Lee.

The Wells in Kent

The Kent Company, which used to supply 30 square miles of the south-east of London as well as 148 square miles of Kent, obtained their water entirely from wells sunk in the chalk beyond the eastern limit of the London Clay. There are also wells at Honor Oak, Streatham, Selhurst and Merton Abbey. In the Kent district there are twenty wells in the chalk and three passing through the chalk into the Greensand. At present the area extending west and east from Woolwich to Southfleet and north and south from the Thames to Westerham is supplied almost wholly with water from these wells. The water requires no filtration and is pumped directly into the mains for immediate supply or to the service reservoirs. In 1920–21 the Kentish wells supplied more than 27,000,000 gallons a day.

Of the chalk wells in Kent belonging to the Water Board the following are a few examples.

A well was sunk near Shortlands station in 1873 to a depth of 250 feet reaching the chalk at a depth of 70 feet. The shaft was carried only 30 feet into the chalk and the remaining 150 feet was bored. The yield at first was 5,000,000 gallons a day.

At Deptford Bridge there are three wells. The chalk here is little more than 20 feet from the surface. One is a shaft well only, carried to a depth of 105 feet. In another the shaft is

95 feet deep and a bore-hole is continued to a total depth of 250 feet. The third has two shafts carried to a depth of 300 feet. The yield of all these wells was tested to 8,000,000 gallons a day.

At Plumstead there are three wells with shafts 135 feet deep. At the bottom of two of these are borings carried to a total depth of 367 feet. At a depth of 118 feet the three shafts are connected by galleries. The chalk is only 6 feet below the surface.

Other Wells in South London

Of the other wells in south London, the well at Honor Oak was made in 1904. The depth is only 300 feet, but the shaft is 11 feet in diameter. It enters the chalk at a depth of 160 feet. At a depth of about 75 feet below the top of the chalk galleries or tunnels have been cut of a total length of 3400 feet so as to make a large collecting area for the water. In these tunnels, or headings, two 8-inch bore-holes have been sunk. The average daily yield of water is nearly 1,000,000 gallons.

The Streatham well was made by the Southwark and Vauxhall Company between 1882 and 1888. Its depth is 1258 feet. It enters the chalk at a depth of 241 feet and a shaft lined with iron cylinders is carried one foot into the chalk. The rest of the well is bored. The bore-hole passes through the chalk, upper greensand and gault and then through beds of the Lower Jurassic period and is continued for 100 feet into the Old Red Sandstone. The yield of water is nearly three-quarters of a million gallons a day. For other wells more than 1000 ft. in depth see p. 63.

At Selhurst, near Croydon, there is a 6-inch boring carried to a depth of 470 feet which gives a small yield of 2250 gallons an hour, but the large well made in 1901 by the Lambeth Company is a shaft 9 feet in diameter carried to a depth of 380 feet. It reached the chalk at a depth of 152 feet, which is 53 feet above sea-level. At the depth of 210 feet headings were cut in the chalk to a total length of 2917 feet, and at a depth of 350 feet there are 950 feet more of headings.

The Merton Abbey well was sunk by the Southwark and Vauxhall Company in 1897. It was finished in 1902. The depth is 332 feet and the chalk was reached at 228 feet. Galleries or headings were made at a depth of 283 feet until a fissure in the chalk was struck which gave an ample supply. The yield is over a million gallons a day.

Property in Underground Water

According to English law the owner of the land is also owner of the water underlying the land. In some cases the owner of the land is owner also of all the minerals underlying the land, but in selling land the minerals are very often reserved. Water, however, goes with the land and the owner may sink a well as deep as he pleases and pump all the water it supplies. This would occasion no inconvenience if the owner could raise only the water which is directly underneath his own land. If a mine-owner takes coal or other mineral from land which is outside his mining rights the law provides a remedy. If a mine-shaft is sunk coal or ironstone will not pass of its own accord from adjoining properties into the shaft to be raised by the miner, but when a well is sunk water will flow into the well from great distances and when it is pumped no one can say whence it comes. It is consequently possible by sinking a deep well and pumping sufficiently to lower the level of the water in other wells in the neighbourhood and perhaps to dry them up entirely if they are shallower than the new well. The greatest, but by no means the only, danger to private wells arises from wells sunk for the purpose of public supply, for these wells are likely to be drawn upon for a very large delivery of water. The people of Hertfordshire naturally complained when Sir Hugh Myddelton carried the water from the Hertfordshire springs to London and the people of Kent were inclined to grumble at the demands made on their chalk water for the London supply. A clause is now inserted in all water Bills requiring that Parliamentary permission should be obtained for any well sunk for public supply. With this restriction water companies and water boards have less powers to take water on their own land than belong to private owners, who can sink wells at their pleasure.

The Early History of London's Water Supply

The London Streams

If we go back in imagination for 1000 years we must picture London as a walled city on the north of the Thames, built mainly upon gravel or brick earth, which rested on clay with the low lands near the river, east, west and south of the City, always liable to floods and generally of a marshy character, and

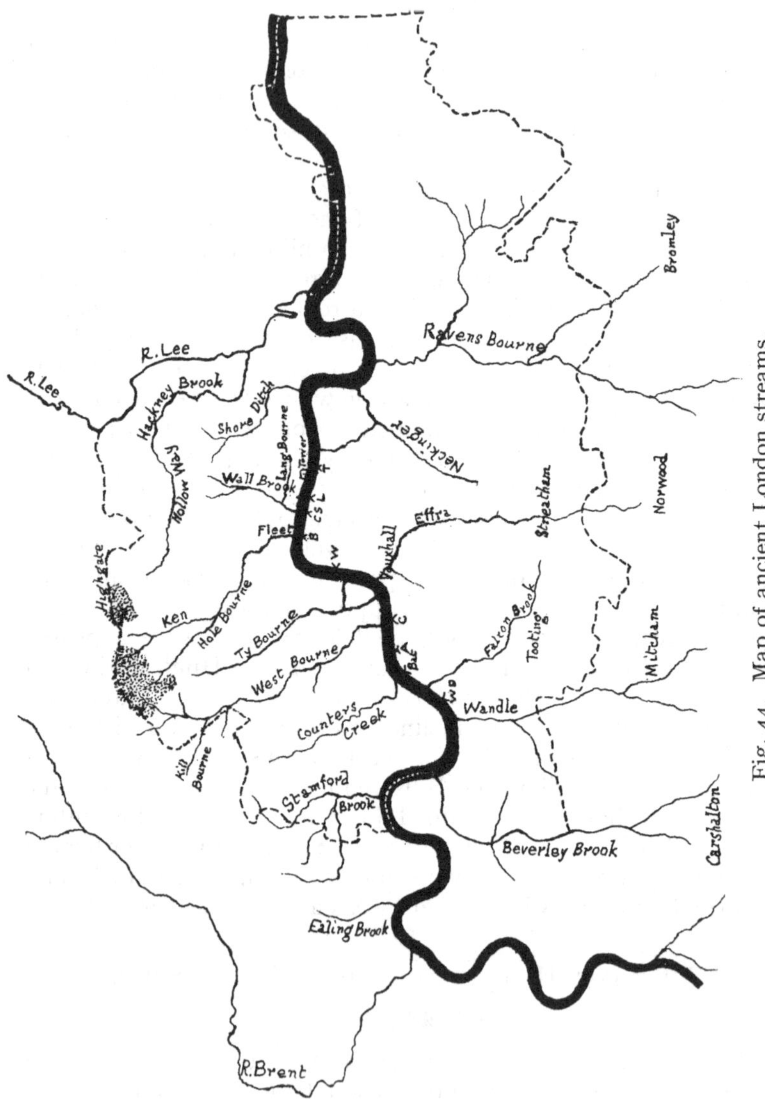

Fig. 44. Map of ancient London streams.

the Hampstead and Highgate ridge separated from the town by three miles or more of open country in which deer, hares and other game were hunted. The Thames was a silver stream abounding in salmon and providing good drinking water for those who could fetch it. Through the City itself flowed the Wall Brook. This stream was outside the Roman city on the west, but flowed through the middle of Saxon London, which extended to Ludgate. A number of small streams flowing from Hoxton and Shoreditch united at Finsbury. The stream so formed passed under the City wall near Liverpool Street Station and is believed to have been navigable from the Thames as far as Moorgate Street. It crossed the Poultry not far from the Mansion House, where it picked up a smaller stream, the Langbourne, which flowed down Lombard Street, and the Wall Brook flowed into the Thames at Dowgate Hill on the west side of Cannon Street Station. Walbrook remains to indicate some part of its course. As in the case of most of the other London streams sewers now take the place of the Wall Brook and its tributaries. Most of the Wall Brook within the City was vaulted over in the latter half of the fifteenth century. Fig. 44 is a sketch-map of the ancient streams. The boundary of the modern county is shown by the broken line except where it is formed by the River Lee on the north-east and by the Beverley Brook on the south-west. The dotted spaces on the northern boundary of the county represent the cap of Bagshot Sands at Hampstead and Highgate.

Just outside the Saxon city but within the modern City of London flowed the Hole Bourne of which the portion extending from Holborn Bridge to the Thames was known as the Fleet River. This river flowed into the Thames opposite De Keyser's Hotel and close to Blackfriars Bridge. The main source of the Fleet was Hampstead Heath on the south-east side of the Spaniards' Road. One stream is said to have flowed down Willow Road to Hampstead Heath Station. Another followed the course of the Hampstead Ponds which were formed by making dams across the valley in the course of the stream. The little river flowed down Fleet Road and keeping to the east of Malden Road picked up the Ken which, starting in Ken Wood (Caen Wood), followed the course of the Highgate Ponds and flowing through Kentish Town joined the Fleet, or Hole Bourne, close to the Regent's Canal between Chalk Farm Road and Kentish

Town Road. The river then flowed along King's Road through Somers Town to Battle Bridge, at King's Cross. Thence it flowed east of Gray's Inn Road along King's Cross Road, in the cutting of the Metropolitan Railway at Farringdon Street Station, and thence down Farringdon Street and New Bridge Street to Blackfriars. It was the valley of the Fleet River which necessitated the construction of Holborn Viaduct. One of Mr Pickwick's early papers, read before the Pickwick Club, was on the Natural History of Sticklebacks as studied by him in the Hampstead Ponds. A short time afterwards he may have fished for the same sticklebacks from his window in the Fleet prison, past the walls of which the river flowed.

But the Fleet was not the only stream which started on Hampstead Heath and found its way to the Thames. There were two others, the Ty Bourne and the West Bourne. The Ty Bourne had its origin at a spring in Shepherd's Fields which was drained when the North London Railway Tunnel was made from Hampstead Heath Station to Finchley Road. The ridge of High Street, Hampstead and Haverstock Hill was the water parting between the Hole Bourne and the Ty Bourne. The stream flowed near Fitzjohn's Avenue to Belsize and past College Villas to Upper Avenue Road. Then it flowed through Regent's Park where the ornamental water has been formed and down Gloucester Place to Oxford Street near Stratford Place. Thence it passed across Lower Brook Street, through Lansdowne Gardens, crossing Piccadilly into the Green Park, and thence to Buckingham Palace. Here it split in two. One part found its way to Great College Street and into the Thames very near the Victoria Tower of the Houses of Parliament while the other crossed the Vauxhall Bridge Road and Grosvenor Road flowing into the Thames west of Vauxhall Bridge. These two arms of the Ty Bourne with the Thames formed Thorney Island. The contractors for the Central London Railway, which passes down Oxford Street, had some trouble with the water when they crossed the now underground course of the Ty Bourne.

The West Bourne started on the west of the Spaniards' Road and the White Stone Pond, not far from Jack Straw's Castle, and crossed the Finchley Road at Heath Drive, picking up another stream from Frognal. Crossing Netherwood Street it was joined by the Kill Bourne, and passing Kilburn Priory it crossed the Edgware Road and having been joined by a tributary from Willesden Lane it flowed by Cambridge Road and Shirland Road and passed Royal Oak Station, Westbourne Terrace and Craven Hill to Hyde Park where the Serpentine was formed on part of its course. Originally the West Bourne fed the Serpentine, but when its water became too much polluted it was diverted into a drain in 1834. From the lower end of the Serpentine the West Bourne passed under Knights' Bridge, thence east of Sloane Street and across Chelsea Bridge Road flowing into the Thames a little to the west of Chelsea Bridge. If you ever travel by the District Railway through Sloane Square Station step on to the platform for a few seconds and above the train you will see the West Bourne crossing the station in an iron pipe about eight feet in diameter supported by a girder at each side, or, to speak more accurately, you will see the iron pipe in which the stream crosses.

Counters Creek joined the Thames at Chelsea Basin after flowing through Paddington, Kensington and Walham Green.

The Hackney Brook flowed through Holloway and Hackney Marshes and joined the River Lee about a mile south of Old Ford. The River Lee remains a navigable river to this day.

In Fig. 44 some of the principal bridges are marked by arrowheads and letters on the south side of the river in order to serve as landmarks. *T* indicates the Tower Bridge; *L*, London Bridge; *CS*, Cannon Street Railway Bridge; *B*, Blackfriars Bridge; *W*, Westminster Bridge; *C*, Chelsea Bridge; *A*, Albert Bridge; *Bat.*, Battersea Bridge; *WD*, Wandsworth Bridge. Waterloo Bridge and Charing Cross Railway Bridge are on the bend of the river between Blackfriars and Westminster Bridges.

On the north side of the Thames in the City and the Strand the land rises somewhat quickly to a height of about 50 feet, and it was this raised terrace overlooking the river which attracted the first settlers and led to the building of a city between the Tower of London and the Wall Brook. The land was dry though surrounded by marsh, and the river provided water and salmon. On the south side the land near the river is at a lower level, and much of Southwark, Bermondsey, and Deptford is so low that even now it is liable to be flooded by an exceptionally high tide in the river. Within the present Administrative County of London there used to flow into the Thames on the south side the Wandle, the Falcon Brook, the Effra, the Neckinger and the

Ravensbourne. The course of the Wandle can be traced on any good map for the river remains an open stream.

The Falcon Brook rose on the south side of Balham Hill and flowed between Clapham Common and Wandsworth Common, across Battersea Rise, down Lavender Hill and into the Thames at Battersea Creek. Falcon Road serves as a reminder of the buried stream.

Until some sixty years ago the Effra flowed as a clear stream from the Norwood Hills through Dulwich and served as a water supply, and there is a tradition that Canute sailed up the Effra as far as Brixton. The stream passed Brockwell Park and near Kennington Oval and joined the Thames a little to the east of Vauxhall Bridge. It is now remembered in Effra Road and Effra Parade.

The Neckinger flowed from Denmark Hill crossing the Old Kent Road to join the Thames near St Saviour's Dock, nearly half a mile east of Tower Bridge. About a mile south of the river it divided and one branch flowed to the east reaching the Thames at Deptford. The Neckinger used to supply water to the Bermondsey Tanneries. There remains Neckinger Road.

The Ravensbourne is still an open stream discharging into the Thames between Deptford and Greenwich.

The Springs and Wells of Old London

In those parts of London which were near the Thames the river afforded the main supply of water for domestic purposes. Away from the river houses were built chiefly where there were beds of gravel or loam. These gravel beds always afforded a good water supply to shallow wells until the increase of population led to the pollution of all the subsoil water in the neighbourhood of the houses. Where the London Clay came to the surface water could not be obtained except in deep wells passing through the clay, and these spots were for long avoided by builders. There were also a number of springs outside the City which flowed with good drinking water. The Anglo-Norman chronicler, Fitz-Stephen, writing between 1180 and 1182, speaks of the water of St Clement's Well as "sweet, wholesome and Cleare." This well was not far from the church of St Clement Danes in the Strand and was very near to Clement's Inn. The spring discharged into a square stone well. In 1756 it had been covered over and a pump fixed for drawing the water. It is generally believed that the well was in the western portion of the site of the New Law Courts. The same Fitz-Stephen refers to the Clerks' Well and the Skinners' Well both in Clerkenwell, a village which grew up around the Priory of St John of Jerusalem and was then separated from the City by fields. The Skinners' Company still own property in Clerkenwell. The Clerks' Well (Fig. 45) was near the west end of Clerkenwell Church (St James'). In 1673 the spring and the plot of ground containing it were given by the third Earl of Northampton for the use of the poor of the parish of St James. There were also a number of other wells fed by springs in Clerkenwell. Another famous well was the Holy Well at Shoreditch given to some Benedictine Nuns for the use of their priory. Holywell Row is still well known to London boys who keep pigeons or other pets. The well was not far from the Curtain Road School.

Fig. 45. The Clerks' Well in 1800.

The great majority of the London wells which were known to the public in the seventeenth, eighteenth and early part of the nineteenth centuries were not used for ordinary domestic supply, but their waters held in solution iron, Epsom salts or other medicinal substances so that they were used as Spa waters, as the waters at Bath, Harrogate, Leamington Spa or Woodhall Spa are used now. The wells were shallow and supplied by springs in the gravel or brick earth lying above the London Clay. They were usually associated with assembly rooms or other places of public entertainment at which refreshments of a different kind were also supplied. As a rule, they were visited in the morning by persons desiring to drink the waters and in the evening by those whose main object was of a more social character. They rose and fell in popularity according to the way in which the house was managed as a place of entertainment. As Clerkenwell, Islington, Hampstead, Battle Bridge, St Pancras, Shadwell, Marylebone, Kensington, Camberwell, Dulwich and other suburbs became thickly populated these shallow springs became too much polluted for drinking purposes, or the excavations for the foundations of houses (and, later on, railway cuttings), but especially the construction of sewers, cut off the supply and the wells were closed and in many cases houses were built over them. The famous Hampstead Chalybeate Well is in the basement of a house in Well Walk, the Sadler's Well is in the basement of a small house in Lloyd's Row, nearly opposite Sadlers Wells Theatre, but as all the wells of this group never formed part of London's ordinary water supply they do not require more than a passing notice here. They are of interest on account of the way in which their memory is, in many cases, preserved in local names.

The Water Carriers

Down to the beginning of the thirteenth century the River Thames and the springs sufficed for London's water supply. The water was brought to the houses by water carriers (commemorated in the signs of the Zodiac by Aquarius), commonly called "Cobs," perhaps because many of them dwelt in a lane known as Cob's Court leading down to the river at Blackfriars. These men would commonly carry two wooden tubs, or "tynes," hanging from a shoulder-yoke (Fig. 46), such as milkmen used to wear, each tyne containing about three gallons of water. Sometimes they carried the water in a tankard on the shoulder. Occasionally it was conveyed in a cart. These water carriers were very numerous and they had their own Guild. About the year 1600 a petition was presented to the House of Commons by the Company of Water-Tankard-Bearers of the City of London and Suburbs in which it was stated that they and their families amounted to 4000 persons. In 1343 the persons living in the streets leading down to the Thames tried to close the streets and exact a toll from every person going to the river to fetch water. Human nature has been ever the same and it is not the great land-owners only who have attempted to extract toll from persons passing through or near their estates.

Fig. 46. The "Cob."

The Conduits

At the beginning of the thirteenth century the increase of population made the existing water supply insufficient. In 1236 Henry III "for the profit of the citie and good of the whole realm thither repairing, for poor to drink, and the rich to dress their meat, granted to the citizens and their successors liberty to convey water from the town of Tybourne (the neighbourhood of Stratford Place, Oxford Street), by pipes of lead into their city." From this time for nearly four hundred years the Corporation of the City of London was regarded as the authority responsible for providing an adequate supply of water. During the period the supply from the river and the wells was supplemented by the construction of twelve conduits bringing water from sources in the suburbs. The conduits were properly the pipes or channels through which the water flowed, but the name was given to the cisterns within the City which were supplied by the conduits and from which the water was drawn through taps, as well as to the buildings which were erected to contain the cisterns. The water carriers brought water from these conduits, as well as from the river, and sold it to the inhabitants. The Tyburn Conduit, constructed in 1237, started at Stratford Place, Oxford Street, past which the Ty Bourne flowed, and terminated in the Great Conduit house in Cheapside, opposite Mercers' Hall. A rate was levied on the inhabitants of Cheapside and the Poultry to meet the working expenses. A second conduit house, known as the Little Conduit, was afterwards erected near St Paul's and supplied from the same source. Though the Corporation was responsible for the water supply, citizens considered the provision of water and the erection of conduit houses to be proper objects of charity, especially in their wills, and by these and other means other conduit houses were erected at Cornhill and elsewhere and their cisterns supplied from the Ty Bourne source. Fig. 47 shows the conduit house at Bayswater. Between 1453 and 1471 under licence from the mayor and corporation the executors of Sir William Eastfield brought water to Fleet Street, Aldermanbury and Cripplegate from Paddington. In 1535 water was brought from Dalston by the City Corporation in pipes to Aldgate where a conduit house was built, and in 1543 Statutory powers were given to the Corporation to bring water from Hampstead. In 1545 a tax was imposed on the inhabitants of the City in order to bring water from some newly discovered springs in Hackney.

Lamb's conduit, still commemorated in Lamb's Conduit Street, was constructed through the private benevolence of William Lamb, a gentleman of the Chapel Royal of Henry VIII. This conduit brought water from Bloomsbury down Lamb's Conduit Street to a conduit house at Snow Hill. The source was not far from the Children's Hospital in Great Ormond Street.

Fig. 47. The Conduit near Bayswater.

There is a conduit house standing in Hyde Park on the North side of Piccadilly just before the end of the railings is reached in going west from Hyde Park Corner. The conduit house is within a few feet of the footway in Piccadilly. Another, much larger and more modern conduit house, is to be seen in Greenwich Park.

Wealthy persons having houses near the routes of the conduits sometimes obtained permission to take a private supply by a pipe or "quill" into their houses and this system so reduced the public supply that it led to many complaints and was one of the causes of the petition of the Tankard Bearers already mentioned. There is a story of a householder in Fleet Street who tapped the conduit without permission and he was punished by being led on horseback round the City conduit houses with a tankard of water on his head which continually leaked through holes made in the bottom and which was filled with water at intervals as required.

The London Bridge Water Works

The time necessarily arrived when, with the increase of London, the conduit supplies from Kensington, Bayswater, Paddington, Marylebone, Bloomsbury, Dalston and Hackney became inadequate. The water carriers could not get from the conduits all the water they wanted for their customers and a special order was issued against the use of weapons in order to secure first place at the taps. Towards the end of the sixteenth century the City Council became anxious about the sufficiency of the supply, but like many modern local authorities they hesitated to incur responsibility in carrying out extensive works under the powers they possessed or under new powers which Parliament was ready to give them, and were glad to transfer their powers to private individuals. Examples of this proceeding were very prevalent in the early days of electric lighting. We have seen how, in some cases, conduits were constructed and conduit houses built by private benefactors with the permission of the City Council. In 1582 a new departure was made and the system of allowing persons to supply water through pipes for private profit, which continued in London until 1904, was commenced. Up to this time the water supply had always flowed from the source to the conduit house by the action of gravity, the conduit house being at a lower level than the source. If water had to be taken uphill from the river it was carried in casks or by water bearers. The lift pump had been invented early in the fifteenth century, but no one had applied it to the purpose of forcing water from the river or from wells through pipes to the houses. When the top of the water in a well was a long way below the surface a bucket suspended by a rope was raised and lowered by a windlass. But in 1580 a Dutchman named Peter Morise applied to the City Council for permission to fix a water-wheel and pump under one of the arches of London Bridge in order to supply water through pipes to the City. He gave a demonstration of the power of his pump by forcing a jet of water over the spire of the Church of St Magnus the Martyr which is close to London Bridge, and much astonished the Lord Mayor, aldermen and councillors. A lease of one arch of the bridge was granted to Peter (whose surname is spelt in a variety of ways) under date May 30th, 1581, for 500 years at a rental of ten shillings a year. In consequence of the success of the undertaking another arch was included in the lease two years later, and afterwards an arch on the Surrey side was similarly used for the supply of water to Southwark. At first the water was forced to a conduit in Leadenhall. In 1582 Bernard Randolph, Common Serjeant of the City, agreed to advance money as a charitable gift "towards bringing water out of the Thames, by an engine to be constructed by Peter Morice, from London Bridge to Old Fish Street, in like manner as he had already brought the water to Leadenhall."

Morice's water-wheels and other machinery were destroyed in the Fire of London and we have no description of them, but they were replaced by his grandson and remained under London Bridge until the bridge was rebuilt in 1825–31. In 1731 a description of the machinery then existing was published in the *Philosophical Transactions of the Royal Society*. In the first arch on the City side was placed a water-wheel carrying twenty-six floats, or paddles, each 14 feet long and 18 inches deep. The wheel was carried on the shorter arms of two levers pivoted on the fixed framework. The longer arms were controlled by chains and could be raised or lowered by a winch. By this means the wheel could be kept so that the floats were just dipping 18 inches in the water as the tide rose and fell. At each end of the axle of the water-wheel was a gear-wheel which drove a pinion. Each pinion turned a four-throw crank-shaft and each crank-pin drove the plungers of two pumps so that there were sixteen pumps driven by this water-wheel.

Fig. 48 shows two of the arches of Old London Bridge with the water-wheels in position but sheltered by a roof.

The third arch had been selected for the extension of this plant and under this arch were placed three water-wheels, one working twelve pumps, another eight and the third sixteen, so that altogether there were fifty-two pumps on the City side of the bridge. The wheels could turn in either direction so that they could be driven by the flowing or ebbing tide and the pumps ceased working only for a short time when the tide was turning. These pumps were designed to force 123,120 gallons of water an hour to a height of 120 feet. A pipe seven inches in diameter was used to convey the water from the pumps.

Fig. 48. London Bridge Waterworks.

The machinery remained the property of Peter Morice's family until 1703 when it was sold to Richard Soams, a member of the Goldsmiths' Company for £38,000. Soams floated a company with a capital of £150,000 and this company also took over the other City conduits for a rental of £700 a year. The water supply of the City was thus handed over by the Corporation to a profit-making company. In 1822 this company was bought by the New River Company.

The New River

Not only the Corporation of the City but the Government of Queen Elizabeth towards the end of the sixteenth century became anxious about the water supply of London, and in 1582 an Act was passed empowering the Corporation to bring water to London within ten years from any part of Middlesex or Hertfordshire. The powers thus conferred on the Corporation were allowed to lapse. In 1605 an Act was passed empowering the Corporation to bring water from the springs of Chadwell and Amwell to London and the scope of this Act was enlarged in the following year, 1606. This Act included not only the springs at Chadwell (Fig. 49) and Amwell but also "other springs in the County of Hertford, not far distant from the same." Still the City Council would not move. The population of London at the time was about 150,000, residing in about 17,000 houses of the type of the old houses still standing at Holborn Bars. Among them there dwelt in Basinghall Street, then called Bassishaw Street, a Welshman, Hugh Myddelton, a citizen and goldsmith, who, like other goldsmiths, carried on also the business of a banker and lived over his shop. He was a friend of Sir Walter Raleigh and the two were occasionally to be seen outside Myddelton's place of business smoking the new weed which Raleigh had brought from America. Hugh Myddelton was born in 1555, one of a family of nine sons and seven daughters. His father was Governor of Denbigh Castle. Hugh was the sixth son. William, the third son, was one of Elizabeth's sea captains. Thomas, the fourth son, was a citizen of London and grocer and became Lord Mayor in 1613. Charles, the fifth son, succeeded his father as Governor of Denbigh Castle. Robert, the seventh son, was a citizen and skinner, and Foulk, the eighth son, was High Sheriff of Denbigh. In 1603 Hugh Myddelton was elected M.P. for Denbigh. His brother Robert was also a member of the House of Commons and the two sat on a Parliamentary Committee entrusted with the consideration of the best method of bringing water to North London. Hugh Myddelton thus became familiar with the details of the problem and when the City Council hesitated to take any action under the Act of 1606 he proposed that the powers of the Corporation should be transferred to him and that he should carry out the work as a private profit adventure. Apparently this suggestion for devolving responsibility was welcomed by the City Council and the powers were transferred to Myddelton on March 28th, 1609, by a Letter of Attorney, Myddelton undertaking to complete the work in

Fig. 49. Chadwell Spring.

four years. Very great difficulty was made by the landowners along the proposed route of the New River. This opposition became so great that in May, 1610, a Bill was introduced in Parliament to repeal the Act of 1606, and on June 20th a committee was appointed to consider the question and report next session. Happily for London's water supply Parliament was prorogued shortly afterwards and did not meet again for four years. But notwithstanding the opposition Myddelton lost little time in getting to work. The first sod was turned at Chadwell, near Ware, on April 21st, 1609. The river was planned to follow approximately the contours of the surface in sections, but with a fall of two inches to the mile. When it was necessary to descend at a more rapid rate a sluice and weir were constructed. The sluice-gates held back the water at the higher level and it poured over the weir to the lower level of the next section. The old Sluice House on the New River at Hornsey was a very favourite object of a Saturday afternoon walk to London schoolboys sixty years ago. The foundation walls of the old sluice are now buried in Wilberforce Road. Although the distance from the New River source to the reservoir at Clerkenwell was only 20 miles, measured in a straight line, the necessity of following the contours of the surface made the total length of the river 38¾ miles. Even with this length it was necessary in some places to cross a valley by means of wooden troughs lined with lead and carried on arches or timber supports, the drainage of the valley flowing underneath these troughs. Fig. 43, p. 102, presents a sketch-map of the New River as it was at the commencement of last century. It will be noticed how the river traverses both sides of the valley in order to cross Pymme's Brook without requiring a bridge of excessive height and length, and that the same course is adopted in crossing other valleys. At the present date the valley between *C* and *D* is crossed by an anti-syphon of cast-iron pipes, shown by a broken line on the map, and the same is the case with most of the other valleys so that the river runs almost directly from *A* to *B* but underground for most of the way. As an example of the distance involved in following contours a study of Fig. 10, p. 27, which represents the high and low level conduits which supplied Jerusalem, is very instructive. The contours could be avoided frequently by lowering the level by means of weirs without the use of pipes, but the former level could not by these means be recovered. According to the Act of Parliament the river was made 10 feet wide and it was generally about 4 feet deep, but the troughs crossing the valleys were 13 feet wide. One of these raised aqueducts at Bush, near Edmonton, was 660 feet in length and the trough was 5 feet deep. Fig. 50 shows the aqueduct over the Madra valley crossing the stream on a brick arch.

When the river had been brought as far as Enfield Myddelton's financial resources appear to have been somewhat strained. The popular story that he died in poverty is not true; for although the New River did not pay a dividend until 1633, after Myddelton's death, Myddelton, who was knighted shortly after the completion of the river and made a baronet in 1622, continued to be a prosperous merchant full of great enterprises

Fig. 50. The timbered New River crossing a valley at Bush Hill.

until his death in 1629. The river passed the grounds of the King's Hunting Lodge at Theobalds, near Enfield, and Myddelton, who had had business transactions with James I in connection with the supply of jewelry for the Queen, may have had other objects in securing the King's interest in his undertaking besides the need of obtaining additional capital to complete the work. At any rate, in November, 1611, an agreement was made between the King and Myddelton under which the King was to provide half the cost of construction and to receive half the profits, but to have no part in the management. Charles I surrendered the King's shares to Myddelton in consideration of a perpetual payment of £500 a year and this is still a tax on the holders of the King's shares in the New River Company. With the financial assistance of the King the river was completed and the opening ceremony took place on Michaelmas Day, 1613, when the water was admitted to the reservoir at the New River Head at Clerkenwell in the presence of the Lord Mayor and Hugh Myddelton's brother, the Lord Mayor-Elect, besides a fashionable City gathering. The Clerkenwell reservoir is about 83 feet above high water. Fig. 51 represents the opening ceremony.

The company was incorporated on June 21st, 1619, under the title of the "Governors and Company of the New River brought from Chadwell and Amwell to London." Sir Hugh had divided his holding into thirty-six adventurers' shares, the King at the time retaining one-half of the capital. These shares as well as the King's shares were afterwards very much subdivided and before the Water Board took over the New River an original share was worth nearly £100,000.

Since Myddelton's time, as indicated above, the course of the New River has been considerably shortened by cutting off the long bends which followed the contours and carrying the water in iron pipes down the hillside, across the valley and up on the other side to join the old course, and the water supply from the springs has been supplemented by water brought by a cut from the upper reaches of the Lee as well as by water from chalk wells in the Lee Valley. The necessity of following the contours at the expense of nearly doubling the length of the river was due to the impossibility in 1609 of obtaining large pipes which could bear the pressure at the bottom of the valleys.

The crossing of the valleys by pipes underground, forming an inverted syphon, instead of following the contours with an open conduit has enabled the New River to be shortened from 38¾ miles to 27 miles. Its average width is now about 24 feet and its depth in the centre varies from 4 to 6 feet. Shortly after leaving its source at Chadwell Spring, above Ware, a basin about 90 feet in diameter into which the spring rises, it is joined by a cut which brings into it water from the Lee. In its course towards Hornsey it receives the water from a number of wells. The total fall in the surface of the water from Chadwell to the New River Head at Clerkenwell is about 18 feet. The water is taken from the river at Hornsey to supply service reservoirs at Crouch Hill, Bourne Hill, Southgate and Hornsey Lane; at Stoke Newington for the reservoirs at Crouch Hill, Bourne Hill,

Fig. 51. "Myddelton's Glory." The admission of the New River water into the Round Pond at Clerkenwell, Sept. 29th, 1613. The original engraving is in the Chairman's room at the Offices of the Metropolitan Water Board.

Finsbury Park, Haggar Lane (Woodford), Maiden Lane and Claremont Square, and at Clerkenwell for the reservoirs at Claremont Square and Crouch Hill. The ancient documents refer to the springs at Chadwell and Amwell, but wells only were mentioned as existing at Amwell when Lord Balfour's Commission was sitting. Originally there was a spring at Amwell, as at Chadwell, but of less importance, and it still exists. Pumping from the wells has deprived the Amwell Spring of most of its water. There are now three wells in Amwell, at Amwell End, Amwell Hill and Amwell Marsh. At the time of going to press, after a phenomenally dry summer and autumn, even the Chadwell Spring is suffering from exhaustion. A few hours' pumping from the wells in the neighbourhood has a considerable effect on the chalk springs and sometimes dries up ditches and streams.

The New River Company still exists, although it has no longer any interest in water supply, for during its long career it accumulated a considerable amount of property in the City and elsewhere which was not required for the purposes of water supply and was not taken over by the Water Board.

The New River was not the first artificial river several miles in length made in this country for the purpose of water supply. In 1591 Sir Francis Drake completed the Leet and presented it to the people of Plymouth. This channel, which brings water from Sheep's Tor on Dartmoor is 24 miles in length, although the distance from end to end in a straight line is only 7 miles. But in 1240 Amicia, Countess of Devon, brought water 5 miles to Tiverton and in 1376 water was brought to Hull from the Aulaby Springs.

The Hampstead Water Company

In 1544 an Act was passed giving powers to the City Corporation to bring water to London from "dyvers great and plentifull springes at Hampsted Hethe, Marybon, Hakkney, Muswell Hill and dyvers places within fyve miles of the saide Citie, very mete, propice and convenyent to be brought and conveyd to the same." Nothing was done under this Act until 1589 when the Hampstead reservoirs were made in the valley between the White Stone Pond at the top of Heath Street and the present Hampstead Heath Station of the North London Railway, on the line of the Hole Bourne. In 1692 the Hampstead Ponds were conveyed to a company known as the Hampstead Water Company, another indication of the desire of the City Council to transfer responsibility for public supply to private hands. Later on the Highgate Ponds were made forming eight reservoirs between Caen Wood and Kentish Town. In 1777 the pond in the Vale of Health was formed (Fig. 52). All these ponds at Hampstead and Highgate were in or near the course of the Hole Bourne,

Fig. 52. The pond in the Hampstead Vale of Health.

or Fleet, or its tributary the Ken. The works of the Hampstead Water Company were taken over by the New River Company in 1855. The water from the ponds is not now used for domestic supply, but it serves well enough for watering roads and flushing drains. In 1920 the ponds supplied nearly 49,000,000 gallons. The water from the ponds flows to a reservoir in Camden Park Road.

Water raising Engines

Reference has already been made to the purchase by the New River Company of the London Bridge Waterworks. In these water power was used for raising the water from the river by means of force pumps. Subsequently a windmill was proposed for increasing the supply of water from a well to one of the conduits, but the proposal was not adopted. Windmills are very commonly used in the country for raising water from wells for private houses and this is almost the only purpose to which windmills are now applied in England. The water supply of London would have been impossible but for the development of the steam-engine, though Dowson gas is now used in the Humphrey pumps at the George V Reservoir at Chingford.

The York Buildings Waterworks

In the Embankment Gardens, between Charing Cross Railway Bridge and the Adelphi Terrace, there is still preserved the old York Gate. York House was situated just above this gate and on part of the site of York House was established the York Buildings Waterworks. The company's powers to supply the West End of London with water from the Thames were originally granted by a patent of Charles II, but in 1691 the company was incorporated by Act of Parliament and, like the New River Company, speculated in lands besides supplying water to its customers. The district supplied included Piccadilly, Whitehall and Covent Garden. In 1818 when the company's mains were purchased by the New River Company there were 2636 householders supplied with water from this source.

The East London Company

The supply in East London north of the Thames, the district of the East London Company, dates from the middle of the seventeenth century. East London had long been very imperfectly supplied with water when Thomas Neale leased land at Shadwell from the Dean of St Paul's and pumped water from the Thames. In 1679 the works were enlarged and in 1687 Neale obtained a Charter for the promotion of a company. This company supplied the district between the Tower of London and Limehouse. In 1808 the East London Company was formed and took over these works and other works which had been established later. They placed their pumping station at Old Ford and made four reservoirs with an area of eleven acres. It was in these reservoirs that the efficiency of storage for purifying water was first discovered by British water engineers. The Roman records are not available. Both the East London Company and the New River Company had powers to supply the East End of London, but by voluntary arrangement in 1815 they defined the area of supply for each company so as not to overlap.

The East London Company bought up the Hackney Waterworks and the Lee Bridge Mills in 1820 and removed their intake from Old Ford to Lee Bridge. The New River Company bought up the London Bridge Waterworks in 1822 but sold the Surrey portion to a private owner. This section of the London Bridge works subsequently led to the formation of the Southwark (and

Vauxhall) Company. The East London Company in 1852 obtained the Act which enabled them to construct the reservoirs at Walthamstow and the filter beds at Lee Bridge and in 1867 they started on the work of bringing water from Sunbury on the Thames right across London to their reservoirs at Walthamstow. The works at Old Ford were finally given up in 1892.

The Kent Company

The Kent Company's supply dates from 1701 when powers were given to the inhabitants of Sayes Court and East Greenwich to take water by pipes from the Ravensbourne. In 1809 the Kent Waterworks Company acquired these rights and in 1811 the right of the Town Commissioners to supply Woolwich. A North Kent Company was established in 1860 to supply Dartford, Crayford, Chislehurst, Bromley, etc., but this company was also absorbed by the Kent Company, and the area of the company's supply extended until, when the Water Board took it over, it amounted to 178 square miles of which 30 square miles were in the County of London. In 1844 a filter bed was made for the Ravensbourne water and a storage reservoir and more filter beds in 1850, but in 1857 the Ravensbourne supply was abandoned in favour of a chalk well. It has already been stated that there are twenty-one of these wells now in the area formerly supplied by the Kent Company and one of them penetrates to the Greensand.

The Recent Companies

The Metropolitan Water Companies which supplied London for some generations before the establishment of the Water Board were

The New River Company, established in 1613.

The East London Waterworks Company, originating in the Shadwell Waterworks of 1669.

The Kent Waterworks Company, starting from the Ravensbourne Waterworks of 1701.

The Chelsea Waterworks Company, established in 1723.

The Southwark and Vauxhall Waterworks Company, dating from 1771.

The Lambeth Waterworks Company, established in 1785.

The Grand Junction Waterworks Company, dating from 1798.

The West Middlesex Waterworks Company, dating from 1806.

These were ordinary commercial companies consisting of shareholders, and Parliament had given to each the right to supply water under certain conditions to a particular area of the London District. Originally it was the opinion of Parliament that a limited amount of competition would be good for the consumer, and the areas of the different companies were made to overlap so that two, or sometimes three, companies would have power to supply water in the same district. The representatives of the different companies would then tout for the order to connect up the houses, and one company would have a main on one side of the street and another company on the other, each having the power to break up the road and make trenches for the pipes. It is said that trench warfare would sometimes be carried on by means of picks, spades and missiles between the navvies employed in the same street by rival companies. A similar overlapping of powers exists to-day in some parts of London between rival electric light companies, but the competition is not carried to physical extremities. After a time the water companies came to the conclusion, at which the nations of Europe have not yet arrived, that it would be to the interest of all for each to have its own exclusive area and keep to it and the limits of the areas were settled by agreement. The competition which Parliament originally provided in the supposed interest of the consumer was thus abolished by mutual arrangement and the supply of water in each district became the monopoly of one company.

The impure character of the water supplied and the high charges made by the monopolising companies led to several enquiries by select committees of the House of Commons between 1820 and 1840. The epidemics of cholera in 1840 and 1849 were held to be due to the impurity of the water and the deficiency in the supply, especially in the poorer and more densely populated districts. The cholera led to the establishment of a General Board of Health which, in 1850, reported on the water supply and recommended that the Thames should be abandoned as a source. There was an intake in the river at Chelsea within a few feet of the mouth of the Westbourne which at that time had become a common sewer. The Home Secretary in 1851 introduced a Bill to transfer the work of the water companies to a public Board. It will be remembered that in 1582 power was given by Parliament to the City Corporation to bring water to London from any part of Middlesex or Hertfordshire, and that the New River was made by Sir Hugh Myddelton under powers granted by Parliament to the Corporation and transferred to him. From that time onwards the water supply of London fell into the hands of private profit companies. There was no local authority outside the City capable of dealing with such a problem. The parish vestries could not provide water for their own parishes separately and the time had not come for the creation of a single municipal authority for Greater London, so when the Home Secretary, Sir George Grey, in 1851 introduced a Bill to place the work of the water companies in the hands of a Special Board, and thus to return to the policy of Parliament adopted 250 years before by placing the water supply in the hands of a public authority, the Bill was defeated through the opposition of the water companies and London had to wait more than 50 years and to pay nearly 40 millions before it could secure a water board.

Parliament was not, however, altogether neglectful of London's water supply during this time. In the next year it passed an Act which required not only that all water taken from the Thames should be taken above Teddington Lock but that all water for domestic use should be filtered except that pumped from deep wells, that all reservoirs within five miles of St Paul's should be covered and that no water should be conveyed in open conduits unless it was afterwards filtered before distribution.

The same Act authorised the Board of Trade to appoint an inspector to enquire into any complaints and required that after five years the water companies should provide a constant supply throughout their districts as soon as every house was furnished with suitable fittings, provided that four-fifths of the owners or occupiers asked for it.

In 1866 a Royal Commission was appointed which enquired into the desirability of obtaining a water supply from Wales, Westmorland or the Severn Valley. The report of this Commission was followed by an Act in 1871 under which a Water Examiner was appointed. For many years analyses of the London water supply were carried out daily in a house in Victoria Street, Westminster.

The Establishment of the Water Board

In 1880 Lord Cross, the Home Secretary, introduced a Bill to buy up the water companies and to place the whole supply in the hands of a Trust, but the Bill came to an untimely end because the Government were thrown out, and afterwards a Select Parliamentary Committee was appointed and recommended that the interests of the water companies should be purchased and the supply entrusted to a public body elected to represent the consumers; but this scheme, which was very nearly carried out, ultimately fell through. In 1889 the London County Council came into existence and during its early years it spent a great deal of time and money in promoting or opposing Bills in Parliament. Whenever a water company applied to Parliament for power to raise more capital or to take more water from the river the County Council was there to oppose the application in order to secure some concession in the interests of the ratepayers. The Council in the meanwhile prepared its own scheme for bringing water from about half a dozen valleys in the Welsh mountains, after the fashion of Liverpool and Birmingham, and this led to the appointment of a Royal Commission in 1892 presided over by Lord Balfour of Burleigh. Reference has already been made to the view taken by this Commission, that London was more favourably situated for water supply than any other great centre of population and it would be wasteful to neglect the water in the London basin in favour of a supply brought from Wales.

In 1897 another Commission was appointed under Lord Llandaff, and this Commission reported in 1899. The County Council made one more attempt in 1901 to obtain an Act enabling it to buy up the water companies and arrange the water supply, but this Bill was opposed by the Government. The main opposition arose on the ground that the Administrative County of London did not represent one-quarter of the total area over which water was supplied by the companies and it was not easy to give to the London County Council the control of the water supply over districts the inhabitants of which had no representation, direct or indirect, on the Council. Some persons were not unwilling to extend the bounds of the Administrative County so as to include the whole of Water London, but Parliament has not even yet seen the wisdom of this step though at the time of going to press this question is prominently before the public.

As against the proposals of the London County Council the Conservative Government of 1901 preferred the creation of a Water Board which should indirectly represent the consumers by including members chosen by the several local authorities. This system was not popular with the Council, which favoured direct representation, and at one time there was considerable danger that the London Education Authority would be appointed on this principle. Carrying out the views of the Government of the time an Act was passed in 1902 constituting the Metropolitan Water Board, which was required to purchase the undertakings of the water companies and to supply the London water area with water. The area entrusted to the Board was not quite the same as that supplied by the companies, for Tottenham and Enfield were added and Richmond, Croydon, Cheshunt and Ware were taken away.

The price to be paid to the companies for their interests in the water supply, including the debts of the companies which had to be taken over, amounted to nearly forty millions sterling. Then the Board had to spend a great deal of money in order to improve the supply so that the total capital outlay in 1921 amounted to more than fifty-three millions. The interest on the capital outlay represented before the war half the cost of the water supplied, or about 4*d*. per thousand gallons. The other half was the cost of pumping, labour, salaries, materials, rates and taxes and general maintenance expenses, including £40,000, now £112,500, a year to the Thames Conservancy and £10,000, now £24,000, a year to the Conservators of the Lee. The total cost was 8·22*d*. per thousand gallons in 1918–19. It is more now.

The amount received by the Water Board for water supplied during the financial year 1920–21 was about £3,085,000, while the interest on capital amounted to £1,714,093. If the Water Board is unable to pay its expenses out of its "water rental" as fixed by its Act it can levy a rate over its district, the rateable value of which is £58,940,044, of which the County and City of London represent £45,597,261.

Originally the Board's rental was a uniform charge of 5 per cent. on the rateable value of the property where the water was supplied for domestic purposes. In 1921 the Board obtained powers to increase this, if necessary, to 8½ per cent. or, with the sanction of the Ministry of Health, to 10 per cent. from 1st

April, 1922, before having recourse to a deficiency rate. The change will give an additional stimulus to the sinking of private wells by large hotels or blocks of commercial offices.

The Metropolitan Water Board consists of sixty-six representatives appointed by the County Councils of London, Essex, Hertfordshire, Kent, Middlesex and Surrey, the City Corporation, the twenty-nine Metropolitan Boroughs, the County Boroughs and other Municipal Boroughs and the Urban District Councils within the water area, the Thames Conservancy and the Lee Conservancy. The London County Council appoints fourteen members, the other County Councils one each, the Metropolitan Boroughs and the larger authorities outside the County one each, making thirty-six, the smaller authorities seven between them and the two conservancies one each, making sixty-six in all. As a rule, the Board is re-elected every third year.

In old times the City Corporation were the Conservators of the Thames up to the City Stone, near Staines, and grants were made by the City Council to companies or individuals permitting them to take water from the river. The water companies used to take their water at Southwark, Battersea and Chelsea long after the Shadwell and the London Bridge Waterworks had been given up, and at that time the London sewers discharged into the river sometimes close to the intakes (see page 127). It was in 1848 that the Lambeth Company removed its intake to Surbiton and in 1852 an Act was passed requiring all supplies to be taken above Teddington Lock.

The conservation of the Thames was entrusted to the Lord Mayor of London in 1687. The powers of the Corporation were transferred to the Thames Conservancy in 1857 and this body was reconstituted by the Port of London Act in 1908, its powers being confined to the portion of the river extending for a distance of about 136 miles to a point 265 yards below Teddington Weir. The lower reaches of the river down to the sea are under the control of the Port of London Authority.

A Departmental Committee, recently appointed by the Ministry of Health to enquire into the water supply of London, reported in July, 1920, on the work of the Water Board that "this great undertaking has been well conducted" and that the wide and populous area has been supplied with a safe and constant supply of water and that elaborate arrangements have been established for keeping the quality under constant and expert supervision.

APPENDIX I

The Cyclone

In this country the principal falls of rain are accompanied by cyclonic movements of the air the centres of which usually travel in an eastward direction so that these disturbances come to us from the Atlantic. It has already been stated that when air ascends it is cooled and if originally saturated with vapour some of this will be condensed into cloud or fog. This is because the air in ascending expands on account of the diminished pressure. In expanding it does work in overcoming the pressure of the air above and around and this work is done at the expense of some of the heat contained in the air. If the temperature remained the same the increased volume could admit more vapour without condensation, but the amount of vapour any space can contain without condensation diminishes very rapidly as the temperature falls, much more rapidly than the volume of the expanding air increases. Condensation therefore takes place. This is why so much rain falls in mountain districts, while on the lee side of an extensive and lofty mountain range there is frequently an arid desert. The air descending from the mountain height contains the latent heat of condensation given up by the condensing vapour, and in addition is warmed again by the work done upon it by the external pressure of the surrounding air as its volume contracts on reaching regions of increased pressure. The air is consequently both hot and dry. The Foehn wind is well known in Switzerland where the air from Italy has left much of its water on the slopes of the Alps.

In a cyclone the warm air charged with vapour does not meet a mountain of rock, but it is constantly sliding up a mountain of very gentle gradients formed by a moving mass of cold air, and as it slides up this atmospheric mountain clouds are formed of the types of nimbus, alto-stratus or cirro-stratus according to their height.

The origin of the cyclone appears to lie in two masses of air of different temperature abutting on one another. If two such masses of air had a vertical plane of contact and the pressure were the same at one level it could not be the same at another. Hence motion must ensue and the cyclone is the dynamical system of stable equilibrium. If the warm and cold air extended to the same height then on the surface of the land the cold air would exert a greater pressure than the warm air and would flow under it in the form of a wedge, the wind tending to blow from the places of higher to those of lower pressure. But as the air moves north or south the rotation of the earth introduces a component into its movements, for the warm air flowing north in the northern hemisphere will have a greater velocity from west to east than the land or the cold air which it passes, while the cold air flowing south will have a less velocity. On this account the wind does not blow, as might otherwise be expected, directly from places of high pressure to places of low pressure at right angles to the isobars, but more nearly parallel to them when observed at the height of the clouds.

In a cyclone the currents of air sweeping past each other produce a rotary movement. Relative to the cyclone the lines of flow are spirals converging towards the centre of the cyclone. In the northern hemisphere the rotation is anti-clockwise as seen by a person looking down on the movements, or it is turning in the direction in which a right-handed screw would be turned to screw it out of a floor. Herr J. Bjerknes of Bergen has given great attention to the cyclones reaching the Norwegian coast from the Atlantic and North Sea, and in 1918 he contributed to the Weather Bureau of the Department of Agriculture of the U.S.A. an account of the structure of moving cyclones which is the basis of the following description.

If the observer places himself at the centre of the cyclone facing in the direction in which the whole disturbance is travelling he will not find that the air is moving round him in circles, as is commonly supposed from the analogy of an ordinary vortex, but all the principal phenomena of the cyclone will lie on his right hand, that is, to the south if the cyclone is moving to the east. In an enlarging sector starting at the centre with a width of about 90° the air will be warm. In the remainder of the horizontal circle the air will be cold and it will be moving in a very different manner in the two regions. The lines separating the warm and cold zones are not straight. The forward line leaves the line of motion of the centre tangentially but gradually curves away to the right. Herr Bjerknes calls it "the steering line" because where it meets the line of motion it is in the same direction. This is the line in which the surface of cold air up which the warm air climbs meets the earth. This surface is not vertical. It is inclined at a very low angle, about 1 in 100, so that to reach a height of a mile and a half the warm air has to travel about 150 miles. The other line leaves the line of motion of the centre at a high angle and curves away to the rear. This line is called the "squall line." Its passage is accompanied by the heavy clearing showers, and the fall of temperature which mark the transit of the cyclone.

The surface of cold air up which the warm air climbs Herr Bjerknes calls the "steering surface." The corresponding surface in the case of the squall line he calls the "squall surface." This inclines towards the rear, but is not a simple inclined plane. It appears to be more of the shape of a breaker.

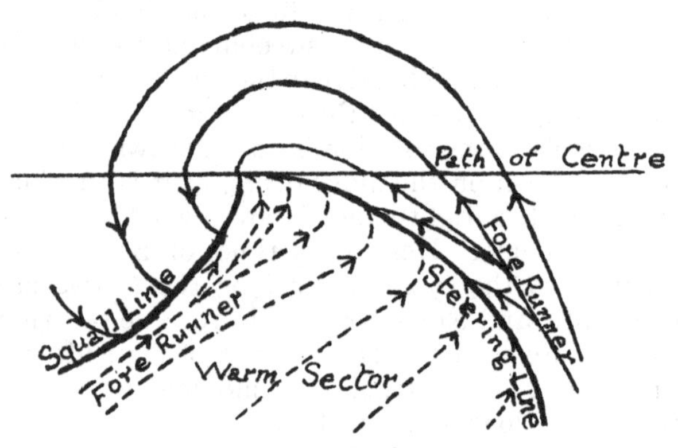

Fig. 53. Cyclone.

Fig. 53 is a reproduction of Herr J. Bjerknes' diagram illustrating his conception of the constitution of a cyclone and showing the lines of flow of the warm and cold air. The streams of warm air are indicated by broken lines.

In places, as shown in Fig. 53, air appears to flow towards both the steering line and the squall line from both sides, warm air from one side and cold air from the other, but the streams do not mix. The warm air flows up the incline of the cold air formed by the steering surface as stated above, while the cold air wedges its way under the warm air at the squall surface. The steering surface and the squall surface meet at the centre forming a "wide flat valley" diverging along curved sides and conveying the warm air up the hillsides over the top of the cold air. This appears to be the way in which the air attempts to attain equilibrium when warm and cold air meet and the balance of pressure is disturbed.

In the figure, which must be regarded as a plan, there are two curved lines called "Fore-runners," one in advance of the steering line and the other in front of the squall line. From these lines air currents appear to flow towards the steering line and squall line respectively, apparently meeting the streams from the other side of those lines, but the streams do not meet at the same level, the warm air passing above the cold.

If we suppose that the climbing slope of the steering surface remains at 1 in 100 to a great height, and if we suppose that the rain-cloud, *nimbus*, is formed at a height of 1600 feet, and may extend to a height of 6000 feet above which alto-stratus may occur and cirro-stratus extend to 25,000 feet, we can foretell what will happen when a cyclone is approaching. About 2,500,000 feet, or nearly 500 miles in front of the centre *cirro-stratus* will make its appearance followed by alto-stratus which will obscure the whole sky. When the centre is within 600,000 feet, or, say, 120 miles, rain will begin to fall through the formation of *nimbus*, gently at first but increasing, becoming heavy as the centre is approached, but on the right of the path of the centre the rain may cease before the steering line passes. During the passage of the warm sector there may be no rain but heavy showers will fall as the squall line passes on account of the steep gradient up which the warm air is being pushed by the advancing wedge of cold air, and the wind will suddenly veer to the right and become very violent. When the squall line has passed there will be a sudden fall of temperature as the cold sector advances and soon the rain will cease, leaving a clear sky as the squall surface, being very steep, quickly passes by.

On the left of the path of the centre an observer will experience the same formation of cloud, but neither the steering line nor the squall line will pass him. The warm air will always be above him and the rain-clouds will be higher than on the other side and continuous until the disturbance has passed and the sky has cleared.

It seems that in accordance with this view of the constitution of a cyclone the axis of the cyclone must lie on the steering surface and therefore point forwards at a gradient of about 1 in 100, so that when the first cirrus appears the cyclone is actually overhead but about 5 miles high, while the point at which the axis of the cyclone reaches the earth is still 500 miles away.

Anyone who has noticed the phenomena attending the passage of a cyclone the centre of which is moving along a line some miles to the north will be struck by the accuracy with which the phenomena observed are foretold from the theory which Herr Bjerknes has advanced.

It must not be supposed that all, or any, cyclones conform precisely to the description which Herr Bjerknes has given. Very much more observation is required, especially of the movements in the higher regions, say above 10,000 feet, before any classification of cyclones can be made, but in the meanwhile it is useful to have in mind a model with which the results of fresh observations can be compared so long as that model is in harmony with the principal phenomena presented in the majority of cases. Herr Bjerknes' description may be taken as that of a type and, like all types, it expresses certain properties of the actual phenomenon by a simplified outline which emphasises some features and ignores others. The special point to which attention should be called is that, like an ocean wave or a sound wave, the cyclone in its forward movement does not carry with it the material the motion of which produces the cyclone[1]. When a wave traverses the ocean the water does not move with the wave. It simply describes a circle or an ellipse, the magnitude of which depends on the height and form of the wave, but is only a few feet in the case of the largest ocean waves and diminishes rapidly as the depth below the surface is increased. It is the system of motion, not the material, which advances with the wave. So in the cyclone, the system of motion advances and may traverse several thousand miles, but as it proceeds fresh quantities of warm air are always being forced up the steering surface incline which is constantly being re-formed by fresh masses of cold air which the cyclone traverses, while in the rear of the centre fresh masses of cold air wedge their way under the warm air below the squall surface, the warm air and the cold air pursuing their respective upward and downward courses in a screw motion which in plan approximates to a logarithmic spiral. With the fresh masses of warm air which take part in the movement fresh supplies of vapour are introduced and fresh clouds are formed some of which descend as rain. Although the

[1] Cf. Shaw and Lampfert. *The Life History of Surface Air Currents.*

clouds appear to be carried with the cyclone along its forward path this is not the case. The clouds follow the movement of the air, not that of the cyclone and fresh clouds are continually being formed along the path of the cyclone and continually falling as rain. It is clear that the same clouds could not continue to pour down rain for several days in succession and over an area several thousand miles in length. The wind, cloud and rain phenomena remain approximately constant and accompany the cyclone, but the materials which exhibit the phenomena are constantly changing, in this respect resembling the organic world wherein the individuals are constantly changing but the type remains for geological periods. A comparison may be made with the Matterhorn cloud described on p. 2. In this case the cloud *appears* to be stationary and clinging to the mountain though the wind is blowing strongly. We *know* that the cloud must be carried by the wind and is therefore being continuously dispersed and re-formed. In this case the mountain is stationary relative to the earth. In the cyclone the cloud system clings to the axis of the cyclone but the axis itself is moving forward. It is the cloud system, not the material of the cloud, which clings.

APPENDIX II

As stated in the Preface, the origin of this little book was a collection of lantern slides, accompanied by memoranda respecting the water supply of London, sent by the Metropolitan Water Board to the London County Council, with the offer to lend the slides at the request of the Local Education Authorities to schools within the Board's area of supply for the illustration of lectures and lessons. Figs. 45, 46, 47, 48, 49, 50 and 52 in this book are direct reproductions from seven of these slides, while a few of the other illustrations have been taken from the same source as other slides or otherwise correspond to them. It was originally intended to print the whole list in this appendix, but since the book was planned the number of slides has increased nearly threefold and the publication of the bare titles and reference numbers of more than 700 slides would increase the dimensions of the book beyond the limits permissible. Moreover, the London County Council is about to supplement its present catalogue of lantern slides for the use of schools by a selection illustrating the Public Services, and in this supplementary catalogue the slides mentioned below will be included. This, perhaps, renders a separate publication of the whole list unnecessary. Teachers in schools within the County of London will know how to obtain the slides from the Council through the Education Officer. Teachers in other schools in "Water London" should apply through their Local Education Authorities to the Metropolitan Water Board, Rosebery Avenue, E.C. 1. There is no subject referred to in this book in connection with the history or present conditions of the Metropolitan supply which is not amply illustrated by the slides in the possession of the Board.

The following is a list of 100 slides selected from more than 700 belonging to the Board which the Education Committee of the L.C.C. are reproducing for use in their schools. In addition they are providing slides corresponding to figures 1, 3, 9, 10, 11, 15, 17, 19, 20, 21, 22, 23, 24, 25, 26, 27, 28, 29, 30, 31, 35, 36, 38, 40, 43 and 44 in this book.

HISTORICAL

1 The Water Carrier or Cob.
2 Old London Bridge.
3 Old London Bridge.
4 Water wheels at London Bridge.
5 Bayswater Conduit.
6 Lamb's Conduit, 1667.
7 Clerks' Well.
8 Sadlers Wells.
9 St Paul's and the Fleet.
10 Entrance to the Fleet.
11 The Fleet, Battle Bridge.
12 Chelsea Water Works, 1750.
13 Green Park Reservoir, 1797.
14 Shepherd's Well, Hampstead.
15 Old Roman Bath in the Strand.

OLD AQUEDUCTS

16 Pont du Gard, Nismes.
17 Claudian Aqueduct.
18 Segovia Aqueduct.

THE NEW RIVER

19 Chadwell Spring.
20 Amwell Marsh Spring.

21 Sir Hugh Myddelton.
22 Map of the New River.
23 New River Head.
24 "Myddelton's Glory"—Opening of the New River.
25 The New River in 1763.
26 London from the New River Head, 1731.
27 The Round Pond, New River Head.
28 Sadlers Wells Theatre and the New River.
29 The New River by Enfield Park.
30 Aqueduct at Bush Hill.
31 Old Sluice Gate, New River.
32 Overmantel in Oak Room, New River Head.

Mains

33 Wooden water mains, method of connection.
34 Wooden pipes.
35 Wooden mains crossing the Fleet.
36 Steel mains to engine house, Staines.
37 Steel syphon tube, Staines.
38 Engine house, Staines, showing delivery mains.
39 Steel breeches pipes, Staines.
40 Covered conduit, Staines.
41 Section of Bayswater Road showing mains.
42 Standpipe tower, Kew Bridge.
43 Shears lifting cap on standpipe, Hampton.
44 Diagram of Venturi meter.

The Thames and Lee

45 Molesey Weir.
46 Chertsey Weir.
47 Diagram of natural flow at Teddington Weir.
48 Map of intakes.
49 Basin of River Lee.
50 Intake, Ferry Lane, River Lee.
51 Intake, Bell Weir, Staines.

Storage Reservoirs

52 Staines Reservoir.
53 Low Maynard Reservoir.
54 River Lee Bridge, Chingford.
55 Chingford intake.
56 Chingford Reservoir, inlet from pumps.

GENERAL

INDEX

Printed in the United States
By Bookmasters